METHODS IN MOLECULAR BIOLOGY

Series Editor
John M. Walker
School of Life Sciences
University of Hertfordshire
Hatfield, Hertfordshire, AL10 9AB, UK

For further volumes:
http://www.springer.com/series/7651

Quantitative Real-Time PCR

Methods and Protocols

Edited by

Roberto Biassoni

Laboratorio Medicina Molecolare, Dipartimento Medicina Traslazionale, Istituto Giannina Gaslini, Genova, Italy

Alessandro Raso

Laboratorio U.O.C Neurochirurgia, Istituto Giannina Gaslini, Genova, Italy

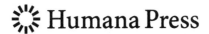
Humana Press

Editors
Roberto Biassoni
Laboratorio Medicina Molecolare
Dipartimento Medicina Traslazionale
Istituto Giannina Gaslini
Genova, Italy

Alessandro Raso
Laboratorio U.O.C Neurochirurgia
Istituto Giannina Gaslini
Genova, Italy

ISSN 1064-3745   ISSN 1940-6029 (electronic)
ISBN 978-1-4939-0732-8   ISBN 978-1-4939-0733-5 (eBook)
DOI 10.1007/978-1-4939-0733-5
Springer New York Heidelberg Dordrecht London

Library of Congress Control Number: 2014936291

Printed on acid-free paper

Humana Press is a brand of Springer
Springer is part of Springer Science+Business Media (www.springer.com)

Preface

From the first report describing real-time PCR detection in 1993, the number of different applications has grown exponentially. Since quantitative PCR is the "gold standard" technology to quantify nucleic acids, thousands of articles and books have been written on both its description and its practical use. Nowadays, it is a very accessible technique, but some pitfalls should be overcome in order to achieve robust and reliable analysis. In this book, our aim is to focus on the different applications of qPCR ranging from microbiological detections (both viral and bacterial) to pathological applications.

Several chapters deal with quality issues which regard the quality of starting material, the knowledge of the minimal information required to both perform an assay and to set the experimental plan. Such issues have been described in the first six chapters, while the others focus on translational medicine applications that are ordered following an approximate logical order of their medical application. The last part of the book gives you an idea of an emerging digital PCR technique that is a unique qPCR approach for measuring nucleic acid, particularly suited for low-level detection and to develop noninvasive diagnosis.

Our hope is that a professional, endowed with the knowledge of some of the methodological issues and of some of the applications, could devise new qPCR-based approaches related to his or her area of investigation. We have tried to cover the possible qPCR methods, but of course we could not cover here all of the feasible applications. We are grateful to all of the colleagues who have contributed to the book with these manuscripts sharing their methods with the qPCR community.

Genova, Italy   *Roberto Biassoni*
                                                                                  Alessandro Raso

Contents

Contributors

JACQUELINE D.H. ANHOLTS • *Department of Immunohematology and Blood Transfusion, Leiden University Medical Center, Leiden, The Netherlands*

ANGELA N. BARRETT • *Department of Obstetrics and Gynaecology, Yong Loo Lin School of Medicine, National University of Singapore, Singapore, Singapore*

ROBERTO BIASSONI • *Laboratorio Medicina Molecolare, Dipartimento Medicina Traslazionale, Istituto Giannina Gaslini, Genova, Italy*

FRANCISCO BIZOUARN • *Gene Expression Division, Bio-Rad Laboratories, San Francisco, CA, USA*

FRANCESCO BROCCOLO • *Department of Health Sciences, University of Milano-Bicocca, Monza, Italy*

STEPHEN BUSTIN • *Postgraduate Medical Institute, Faculty of Health, Social Care and Education, Anglia Ruskin University, Chelmsford, Essex, UK*

LYN S. CHITTY • *Clinical and Molecular Genetics Unit, UCL Institute of Child Health, London, UK; North East Thames Regional Genetics Laboratory, Great Ormond Street Hospital, London, UK*

FRANS H.J. CLAAS • *Department of Immunohematology and Blood Transfusion, Leiden University Medical Center, Leiden, The Netherlands*

MICHAEL EIKMANS • *Department of Immunohematology and Blood Transfusion, Leiden University Medical Center, Leiden, The Netherlands*

BERND FALTIN • *Applied Research 1, Microsystem Technologies-Microstructuring and Assembly, Robert Bosch GmbH, Stuttgart, Germany*

LETIZIA FORONI • *Imperial Molecular Pathology Laboratory, Hammersmith Hospital Campus, Imperial College London, London, UK*

PIERRE FOSKETT • *Imperial Molecular Pathology Laboratory, Hammersmith Hospital Campus, Imperial College London, London, UK*

GARETH GERRARD • *Imperial Molecular Pathology Laboratory, Hammersmith Hospital Campus, Imperial College London, London, UK*

JAN HELLEMANS • *Biogazelle, Zwijnaarde, Belgium*

FRANK HUFERT • *Department of Virology, University Medical Center Göttingen, Göttingen, Germany*

JIM HUGGETT • *Molecular and Cell Biology, LGC Ltd, Teddington, UK*

GEMMA JOHNSON • *Blizard Institute of Cellular and Molecular Science, Queen Mary University, London, UK*

BENEDIKT KIRCHNER • *Physiology Weihenstephan, ZIEL Research Center for Nutrition and Food Sciences, Technische Universität München, Freising, Germany*

MICHAEL LEHNERT • *Laboratory for MEMS Applications, IMTEK – Department of Microsystems Engineering, University of Freiburg, Freiburg, Germany*

MAURO SEVERO MALNATI • *Human Virology Unit, Division of Immunology, Transplantation and Infectiuos Diseases, Fondazione Centro San Raffaele, Milan, Italy*

EDDI DI MARCO • *UOC Laboratorio Centrale di Analisi, Istituto Giannina Gaslini, Genova, Italy*

SAMANTHA MASCELLI • *UOC Neurochirurgia, Istituto Giannina Gaslini, Genova, Italy*

SWANHILD U. MEYER • *Physiology Weihenstephan, ZIEL Research Center for Nutrition and Food Sciences, Technische Universität München, Freising, Germany*

TANIA NOLAN • *Sigma-Aldrich, Haverhill, Suffolk, UK*

AFIF ABDEL NOUR • *Special Infectious Agents Unit-Biosafety Level 3, King Fahd Medical Research Center, King Abdulaziz University, Jeddah, Saudi Arabia*

CLAUDIO ORLANDO • *Department of Biomedical, Experimental and Clinical Sciences, University of Florence, Florence, Italy*

VIJAY PAUL • *Physiology Weihenstephan, ZIEL Research Center for Nutrition and Food Sciences, Technische Universität München, Freising, Germany; National Research Centre on Yak (ICAR), Dirang, Arunachal Pradesh, India*

MARIO PAZZAGLI • *Department of Biomedical, Experimental and Clinical Sciences, Clinical Biochemistry Unit, University of Florence, Florence, Italy*

MICHAEL W. PFAFFL • *Physiology Weihenstephan, ZIEL Research Center for Nutrition and Food Sciences, Technische Universität München, Freising, Germany*

PAMELA PINZANI • *Department of Biomedical, Experimental and Clinical Sciences, University of Florence, Florence, Italy*

ALESSANDRO RASO • *Laboratorio U.O.C Neurochirurgia, Istituto Giannina Gaslini, Genova, Italy*

IRMGARD RIEDMAIER • *Physiology Weihenstephan, ZIEL Research Center for Nutrition and Food Sciences, Technische Universität München, Freising, Germany*

STEFANIE RUBENWOLF • *Laboratory for MEMS Applications, IMTEK – Department of Microsystems Engineering, University of Freiburg, Freiburg, Germany*

DOMENICO RUSSO • *Human Virology Unit, Division of Immunology, Transplantation and Infectiuos Diseases, Fondazione Centro San Raffaele, Milan, Italy*

FRANCESCA SALVIANTI • *Department of Biomedical, Experimental and Clinical Sciences, University of Florence, Florence, Italy*

STEFFEN SASS • *Institute of Computational Biology, Helmholtz Center Munich, German Research Center for Environmental Health, Oberschleißheim, Germany*

FELIX VON STETTEN • *Laboratory for MEMS Applications, IMTEK – Department of Microsystems Engineering, University of Freiburg, Freiburg, Germany; Institut für Mikro- und Informationstechnik, HSG-IMIT, Freiburg, Germany*

KATHARINA STOECKER • *Physiology Weihenstephan, ZIEL Research Center for Nutrition and Food Sciences, Technische Universität München, Freising, Germany*

FABIAN J. THEIS • *Institute of Computational Biology, Helmholtz Center Munich, German Research Center for Environmental Health, Oberschleißheim, Germany; Department of Mathematics, Technische Universität München, Garching, Germany*

ELISABETTA UGOLOTTI • *Dipartimento di Medicina Sperimentale, Istituto Giannina Gaslini, Genova, Italy*

JO VANDESOMPELE • *Biogazelle, Zwijnaarde, Belgium; Center for Medical Genetics, Ghent University, Ghent, Belgium*

IRENE VANNI • *Dipartimento di Medicina Sperimentale, Istituto Giannina Gaslini, Genova, Italy*

SIMON WADLE • *Laboratory for MEMS Applications, IMTEK – Department of Microsystems Engineering, University of Freiburg, Freiburg, Germany*

MANFRED WEIDMANN • *FK9 4LA Stirling, Scotland, UK*

ROLAND ZENGERLE • *Laboratory for MEMS Applications, IMTEK – Department of Microsystems Engineering, University of Freiburg, Freiburg, Germany; Institut für Mikro- und Informationstechnik, HSG-IMIT, Freiburg, Germany; BIOSS – Centre for Biological Signaling Studies, University of Freiburg, Freiburg, Germany*

Chapter 1

Twenty Years of qPCR: A Mature Technology?

Alessandro Raso and Roberto Biassoni

Abstract

Quantitative PCR is the "gold standard" technology to quantify nucleic acids and, since the first report describing real-time PCR detection in 1993, its use has been grown exponentially. More recent technological advancements have extended the field of applications ranging from high-resolution melting detection to digital PCR. Nowadays, it is a very accessible technique, but some pitfalls should be overcome in order to achieve robust and reliable analysis.

Key words qPCR, HRM, dPCR, Dye-labeled probe, Intercalating dye

Quantitative real-time PCR (qPCR) [1] is a sensitive and robust technique directly evolved from the "end-point detection" polymerase chain reaction (PCR) [2]. PCR is a polymerase-dependent repetitive thermal reaction able to generate copies of a specific template spanning from two short oligo-deoxynucleotide sequences (primers). The qPCR is at least 100-fold more sensitive than the end-point detection and displays a dynamic range not less than nine logs.

In general, PCR is based on the quantitative relationship between the amount of target sequence at the beginning and the amount of amplified PCR product at any given cycle. Such correlation follows an exponential rate that gives rise to an exact doubling of product that is accumulated at every cycle (when 100 % reaction efficiency is assumed). Such exponential phase is limited to a short number of PCR cycles since the reaction occurs in a classical closed system. This state causes the depletion of reactants concentrations, enzyme activity, and other factors, while the products accumulate over time. Thus, the PCR is characterized by four reaction phases known as: Baseline, Exponential, Linear, and Plateau. Baseline is very short step where the amplification is not yet detectable. During the second phase of amplification the kinetic of reaction determines a favorable doubling of amplicons. Linear phase is characterized by slowdown trend of amplification and the products

Roberto Biassoni and Alessandro Raso (eds.), *Quantitative Real-Time PCR: Methods and Protocols*, Methods in Molecular Biology, vol. 1160, DOI 10.1007/978-1-4939-0733-5_1, © Springer Science+Business Media New York 2014

are no longer doubled at each cycle. Finally, at the plateau the reaction is essentially terminated, no more accumulation of amplicons is achieved even if the number of cycles is increased and, very unuseful, PCR products may start to degrade. Traditional PCR detects reaction products at such last phase, thus it is so called "end-point PCR." Worthily, each reaction can reach the plateau at a different point and may be characterized by different kinetics that leads to different performances. Thus, end-point PCR cannot be used for quantification purposes. Contrary to the end-point PCR, the qPCR allows the immediate detection of amplified products at any given cycle using a quantitative relationship with the target sequence at the beginning of reaction.

Real-time detection should be performed during the exponential phase where the fluorescence signal is directly proportional to DNA concentration.

Two essentially different types of chemical strategies ensure such generation of fluorescent signal. One is based on double-stranded intercalating dye (SYBR-Green and its evolution) and the other can use a *plethora* of different dye-labeled probe systems (i.e., exonuclease-based double-labeled dye oligo-deoxynucleotide, molecular beacons) [3, 4]. In general, qPCR detection achieved using intercalating dye is defined as "nonspecific," while detection by fluorescent probes is considered "template-specific" [3, 5]. Such dichotomy assumes that the use of probe introduces an additional level of specificity since it does not produce any fluorescence signal, due to probe hybridization, for amplicons generated by either mis-priming or primer-dimers.

On the implementation of qPCR point of view a new powerful technique called HRM (high-resolution melting) has also been developed. Such technique is a postamplification analysis that uses data acquired at the plateau phase in order to accurately quantify mutations, polymorphisms, and epigenetic differences on double-stranded nucleic acid molecules. It is based on the use of both saturating dye and instruments that are able to monitor tiny differences of melting temperature of the double strand. HRM is an excellent alternative to the classic molecular methods for screening genetic variants such as dHPLC sequencing.

Thus, it confirms continuous improvements of qPCR along its two decades of life leading to novel technical evolutions.

While the number of its applications is increasing exponentially as the mechanism of PCR itself, there is neither consensus on experiment design nor homogeneity in practice [6]. Therefore, in order to achieve reliable experiments and unequivocal interpretation of qPCR data, several practical guidelines have been recently proposed [6–8].

The third generation of the PCR is the digital PCR (dPCR) that should even be considered a modified qPCR showing high sensitivity that allows an absolute quantitation. In fact, it is a hybrid application using a classic PCR reaction together with fluorescence-based detection. Both the extreme dilution and partitioning of the sample yield to produce single-molecule subreactions, some of which carry target sequence while others do not; such ratio is used to quantify the starting amount of the target template.

Therefore, the wide range of possible applications of qPCR, such as clinical diagnosis, molecular research, and forensic studies, already make it a mature but not outdated technology.

References

1. Higuchi R, Fockler C, Dollinger G et al (1993) Kinetic PCR analysis: real-time monitoring of DNA amplification reactions. Biotechnology (NY) 11:1026–1030

2. Mullis KF, Faloona S, Scharf R et al (1986) Specific enzymatic amplification of DNA in vitro: the polymerase chain reaction. Cold Spring Harbor Symp Quant Biol 51:263–273

3. Bustin SA, Nolan T (2004) Pitfalls of quantitative real-time reversetranscription polymerase chain reaction. Biomol Tech 15:155–166

4. Ishiguro T, Saitoh J, Yawata H et al (1995) Homogeneous quantitative assay of hepatitis C virus RNA by polymerase chain reaction in the presence of a fluorescent intercalater. Anal Biochem 229:207–213

5. Bonnet G, Tyagi S, Libchaber A et al (1999) Thermodynamic basis of the enhanced specificity of structured DNA probes. PNAS USA 96:6171–6176

6. Bustin SA, Benes V, Garson JA et al (2009) The MIQE guidelines: minimum information for publication of quantitative real-time PCR experiments. Clin Chem 55:611–622

7. Raymaekers M, Smets R, Maes B et al (2009) Checklist for optimization and validation of real-time PCR assays. J Clin Lab Anal 23:145–151

8. Taylor S, Wakem M, Dijkman G et al (2010) A practical approach to RT-qPCR-Publishing data that conform to the MIQE guidelines. Methods 50:S1–S5

Chapter 2

Minimum Information Necessary for Quantitative Real-Time PCR Experiments

Gemma Johnson, Afif Abdel Nour, Tania Nolan, Jim Huggett, and Stephen Bustin

Abstract

The MIQE (minimum information for the publication of quantitative real-time PCR) guidelines were published in 2009 with the twin aims of providing a blueprint for good real-time quantitative polymerase chain reaction (qPCR) assay design and encouraging the comprehensive reporting of qPCR protocols. It had become increasingly clear that variable pre-assay conditions, poor assay design, and incorrect data analysis were leading to the routine publication of data that were often inconsistent, inaccurate, and wrong. The problem was exacerbated by a lack of transparency of reporting, with the details of technical information inadequate for the purpose of assessing the validity of published qPCR data. This had, and continues to have serious implications for basic research, reducing the potential for translating findings into valuable applications and potentially devastating consequences for clinical practice. Today, the rationale underlying the MIQE guidelines has become widely accepted, with more than 2,200 citations by March 2014 and editorials in Nature and related publications acknowledging the enormity of the problem. However, the problem we now face is rather serious: thousands of publications that report suspect data are populating and corrupting the peer-reviewed scientific literature. It will be some time before the many contradictions apparent in every area of the life sciences are corrected.

Key words PCR, Reverse transcription, Diagnostics, Gene expression

1 Introduction

The MIQE (minimum information for the publication of quantitative real-time PCR) guidelines [1] represent a major milestone in the transformation of the real-time quantitative polymerase chain reaction (qPCR) from a research technique into a reliable "gold standard." A comparison of qPCR with conventional endpoint PCR reveals that qPCR is less prone to contamination, easier to implement, requires less hands-on time, has the potential for high throughput, and can be quantitative. This has allowed it to rapidly displace legacy PCR for many applications, making it into a ubiquitous technique capable of delivering numerous results in minimal

Roberto Biassoni and Alessandro Raso (eds.), *Quantitative Real-Time PCR: Methods and Protocols*, Methods in Molecular Biology, vol. 1160, DOI 10.1007/978-1-4939-0733-5_2, © Springer Science+Business Media New York 2014

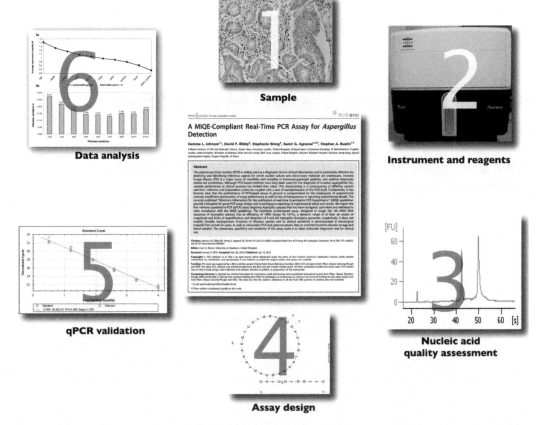

Fig. 1 A qPCR publication depends on the successful completion of a series of steps, each one of which must be carefully quality controlled to ensure reliable, accurate, and reproducible amplification

time. The simplicity of data acquisition has fostered the impression that qPCR data are robust and reliable, but has failed to impress on researchers that there are numerous critical steps associated with a successful qPCR assay, every one of which needs to be quality controlled for the results to be meaningful (Fig. 1). Unfortunately, it has been clear for some time that the quantity of qPCR data is not matched by an equivalent quality. As a consequence, there are numerous publications reporting contradictory data and results are frequently not reproducible, yet are circulated in the peer-reviewed literature without any obvious criteria to distinguish a genuine result from a technical artifact.

A particular low point came with the revelations concerning the inappropriate use of the reverse transcription (RT)-qPCR in publications associating measles virus with novel gut pathology and autism [2]. A public dissection of published data at the Washington DC autism trial in 1997 revealed a catalogue of inconsistencies, including the use of inappropriate samples, protocols, and analysis methods, ignoring of negative controls that were positive and amplification of DNA contaminants. This resulted in a conclusion "confirming" an association between the presence of measles virus and gut pathology in children with developmental disorder. The results were widely used to link the measles, mumps, and Rubella (MMR) vaccine to the occurrence of autism in children and provide a graphical example of the damage an improperly conducted and inadequately published scientific study can do. Other, less egregious examples of the problems that arise from poor experimental practice include the controversy surrounding the lack of association of xenotropic murine leukaemia virus-related virus (XMRV) in prostate cancer [3] and chronic fatigue syndrome [4] and the retraction of a paper describing the migration of mRNA to initiate flowering, which was a "breakthrough of the year" [5].

In response, a growing consensus has been developing around the need to improve the transparency of reporting of relevant experimental detail to include every aspect important to the qPCR assay itself as well as issues relating to pre- and post-assay parameters. Specifically, it became clear that there is a requirement for a set of recommendations that can be used by journal reviewers, who need to be able to evaluate the reliability of the experimental protocols and ensure the inclusion of all essential information in the final publication. Whilst there had been numerous individual papers highlighting the inadequacies, misconceptions, and failures of this important and ubiquitous enabling technology (reviewed in ref. 6), there had been no unifying proposals for a solution to these problems. This need was addressed by the publication of the MIQE guidelines, coauthored by an international group of researchers with a long history of involvement in addressing quality-related issues. For the first time there was a focus that enabled other researchers, journal editors, and non-qPCR expert readers of publications to understand what to look for when evaluating the reliability of conclusions derived from publications utilizing qPCR-based technologies.

There has been a rapid expansion in the number of researchers aware of the existence of these guidelines as well as an increasing number of citations of the original publication in the peer-reviewed literature (> 2,200 by March 2014). There even is an iOS/Android app available for mobile telephones and tablet computers [7]. The final acceptance of the need for guidelines such as these was an editorial in Nature, published in April 2013, which acknowledged that "journals such as this one compound them [the problems] when they fail to exert sufficient scrutiny over the results that they publish" and called for a "checklist" that "focuses on a few experimental and analytical design elements that are crucial for the interpretation of research results but are often reported incompletely" [8]. As a result, Nature and its associated journals no longer have space restrictions on the methods section and even though this conversion by Nature is very late, it is nonetheless welcome.

It is certainly a long way from the 2010 Nature Medicine report on MIQE, which quoted its editor's attitude as "We would be delighted to embrace the [MIQE] guidelines, but we are not really persuaded that the guidelines are embraced by the community" [9]. It is unfortunate that there was no sign of leadership from the high impact factor journals then, as their support would have accelerated the acceptance of the guidelines.

2   The Guidelines

The MIQE guidelines offer a strategy for reproducibility and quality control that allows scientists to cultivate better practices in quantitative PCR experiments [10]. Their fundamental goal is to encourage the publication of transparent and comprehensive technical detail, since this allows a reader to take technical excellence for granted and to focus on the biological relevance of that publication's conclusions. A corollary is that they include all the information required to design, validate, and optimize an assay from scratch and so constitute a blueprint for good assay design. Anyone using MIQE as the basis for developing a qPCR-based assay is virtually guaranteed to achieve that goal and obtain an efficient, specific, and sensitive assay.

MIQE consists of nine sections, with 85 parameters that constitute the minimum information required to allow potential reproduction as well as unambiguous quality assessment of a qPCR-based experiment. These nine sections comprise

- Experimental design
- Sample properties
- Nucleic acid extraction and quality assessment
- Reverse transcription
- Target information
- Primer and probe details
- qPCR protocol optimization and validation details
- Data analysis

The 85 parameters fall into two categories: some are deemed to be essential and are labeled "E" in the published guidelines, because they are indispensable for an adequate description of the qPCR assay. Other components are more peripheral and are labeled "D" (desirable), yet represent an effective foundation for the implementation of best practice protocols. Adherence to these parameters also encourages much-needed standardization, especially important when using qPCR assays for diagnostic applications. Importantly, these parameters are based on common sense and current best practice and so are not set in stone and remain open for discussion; indeed, a slightly modified version, labeled MIQE précis encompasses the key MIQE parameters essential for publication in Biomed Central (BMC) journals [11]. Most recently, MIQE-style guidelines for minimum information for publication of quantitative digital PCR experiments (dMIQE) have been published [12].

Possibly the most contentious part of the original MIQE guidelines was the essential requirement for publications to report the sequences of any primers used and the suggestion to also report the sequences of any probes. The rationale behind this is rather straightforward: an experiment cannot be reproduced exactly if the primer sequence, one of the principal reagents, is unavailable. Lack of access to a probe sequence, on the other hand, does not preclude analysis of the specificity, efficiency, and sensitivity of an assay; however, for completeness' sake it is but a small step to take for most researchers. Many commercial qPCR assays are not supplied with the primer/probe sequences, since most vendors consider this commercially sensitive information; usually there are also no details provided on empirical validation of each individual assay. Publications utilizing such assays could not satisfy the original MIQE requirements, placing limits on a universal acceptance of MIQE.

Consequently, an amendment of the original guidelines now requires either primer sequences or a clearly defined amplicon context sequence [13]. This guidance was issued based on the assessment that in the absence of full primer sequence disclosure it is possible to achieve an adequate level of transparency, but only if there is an appropriate level of background information and disclosure of validation results on the qPCR assay. Consequently, if primer sequences are not disclosed, a MIQE-compliant publication should institute the same validation criteria used for assays reporting primer/probe sequences. Specifically, when reporting a precise fold-change for a transcript it remains an essential requirement that the PCR efficiency, analytical sensitivity, and specificity of each individual assay be determined. This information should be verified by the investigator for the actual assay that is being reported using the conditions and personnel in their laboratory and not extrapolated from commercial assays validated by the vendors.

It is worth emphasizing that MIQE proposes minimum guidelines; hence more information can be disclosed, if so desired. For example, MIQE requests information about the specificity, PCR efficiency, r^2 of calibration curves, linear dynamic range, and C_q variation at the limit of detection. Including the data in a table can fulfill these requirements. However, the addition of individual calibration and melt curves in supplementary material would be far more informative and allow the reader to get a much better feel for the quality of the published data.

3 Why the Need for Such Detail?

At first sight the requirement to list 85 individual criteria appears to be rather onerous. However, every one of the parameters is likely to be encountered and addressed during the routine development, optimization, and validation of a qPCR assay. Hence it is usually simply a matter of recording the results, which can then be tabulated and submitted with the manuscript. The use of the MIQE app also simplifies MIQE compliance, as the analyzed data can be exported with a single click and can then be attached as supplemental data to the article.

The information requested for the reverse transcription step provides a handy example of why the guidelines incorporate such detailed criteria. They list five essential (complete reaction conditions, amount of RNA and reaction volume, priming oligonucleotide if using gene-specific priming and concentration, temperature, and time) and three desirable (manufacturer of reagents and catalogue numbers, C_qs with and without RT, storage conditions of cDNA) parameters. The reason for this is that RT yields depend on total RNA concentration and RT reaction conditions such as the priming strategy, which affects RT efficiency and is different for different target genes [14]. This is demonstrated in Fig. 2, where the C_qs of various target mRNAs differ according to whether cDNA synthesis was primed by random hexamers, pentadecamers, oligo-dT, or gene-specific primers. Assay details are shown in Table 1. Since it cannot be predicted how different priming methods affect the RT efficiency of each target, it is essential that a detailed description of the protocol and reagents used to convert RNA into cDNA be provided. Furthermore, reverse transcription yields can vary significantly with the choice of reverse transcriptase and, as with the priming strategy, this variation is gene dependent [15]. This variability is demonstrated in Fig. 3, with the maximum ΔC_q recorded by different RTs ranging from 4.5 (22-fold) and 7 (128-fold), depending on the target.

Quality control of nucleic acids is another example of the detailed reporting suggestions proffered by the MIQE guidelines. Whilst most researchers are aware of the importance of measuring RNA integrity prior to quantification, many fail to ensure adequate purity of their samples. Purity does not refer to a sample's A_{260}/A_{280} ratio, but rather encompasses the absence of inhibitors of either the RT or the PCR reaction. Inhibition is a well-known yet poorly described phenomenon and we were the first to propose a universal method for inhibition testing that involves the use of a template expressed only in potatoes [16]. The technique, called SPUD, compares the C_qs obtained from the amplification of SPUD templates suspended in water with those obtained from SPUD templates spiked into sample preparations. The example in Fig. 4a shows the huge range of C_qs obtained with samples extracted from

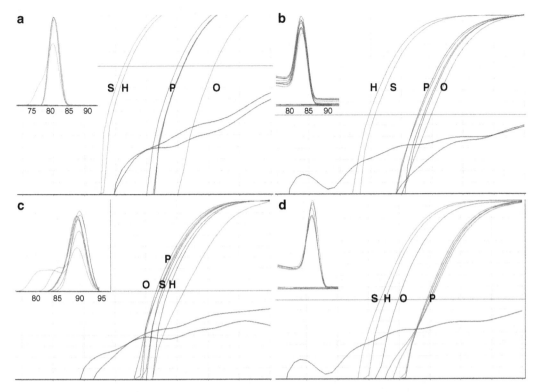

Fig. 2 Effect of priming strategy on final C_q. Equal amounts of RNA (RIN = 10) were reverse transcribed using gene-specific primers (S), random hexamers (H), oligo-dT (O), or pentadecamers (P) at different concentrations, (**a**) DRG-1 (**b**) p21 (**c**) E-cadherin (**d**) Osteopontin. The *inserts* show the respective melt curves. All assays were carried out on a Corbett 6000 qPCR instrument (95 °C, 10 s; 59 °C or 60 °C, 15 s; 72 °C, 30 s) × 40

Table 1
Details of primers and amplicons

Accession no.	Name	Primers	Ta (°C)	Efficiency (%)	Amplicon size (bp)	Position (start)
NM_006096	DRG-1	CGATTTGCTCTAAACAACCCTGAG CATCCAGCCTTCCGCACAAG	58	100	78	582
NM_000582	Osteopontin	TTAAACAGGCTGATTCTGGAAGTTC GATTCTGCTTCTGAGATGGGTCA	60	99	105	221
NM_000389	p21	CTGGAGACTCTCAGGGTCGAA GGATTAGGGCTTCCTCTTGGA	60	99	98	523
NM_004360	E-cadherin	TCCTCAGAGTCAGACAAAGACCAG TCCTCGCCGCCTCCGTAC	59	100	95	2,672
NM_001168	Survivin	CAGTGTTTCTTCTGCTTCAAGGAG AGCGCAACCGGACGAATG	62	98	90	287
NM_002467	c-myc	TGAGGAGACACCGCCCAC CAACATCGATTTCTTCCTCATCTTC	62	100	71	1,292

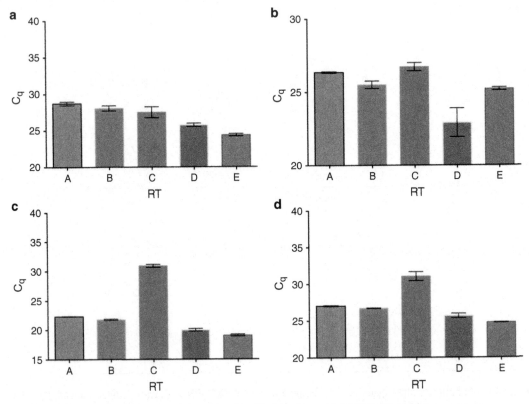

Fig. 3 Different properties of five different RTs. (**a**) p53; (**b**) c-myc; (**c**) GAPDH; (**d**) survivin

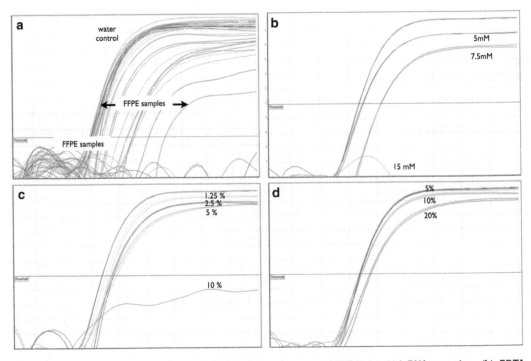

Fig. 4 Inhibition of qPCR assays using SPUD as a reporter. (**a**) FFPE extracted RNA samples; (**b**) EDTA; (**c**) Phenol; (**d**) Ethanol

formalin fixed, paraffin-embedded samples, indicative of significant inhibition of most samples. Figure 4b–d shows the effects of three different common inhibitors of the PCR reaction on the SPUD assay. An interesting, and so far unexplained observation is that the slopes of the reactions remain rather similar, whereas the C_qs increase with increasing concentrations of inhibitor. Whilst the SPUD assay is useful for general detection of inhibition, it has become clear that inhibition is not a simple process that affects each and every template to the same degree. Rather, testing for inhibition of individual targets within a sample is important, as we have demonstrated that inhibitors affect different PCR reactions to different extents [17].

The question of how to normalize appropriately when measuring RNA levels is one that has been around since the early days of RT-qPCR [18] and continues to be dealt with in a wholly unsatisfactory manner [19]. The problem is that in addition to problems associated with reverse transcriptases, priming methods, inhibitors, and PCR efficiencies there is an inherent variability, i.e., error associated with RNA itself and with the protocols used for its extraction. This requires the application of a consistent, appropriate, and accurate method of normalization to control for that error. There are several normalization strategies, none of which are mutually exclusive and all of which can be incorporated into a protocol at many stages [20]. For now, the use of reference genes represents the strategy that is most widely accepted, but they must be validated within the context of each individual experimental setup if the data are to be biologically meaningful [21, 22]. As a general rule, if RNA levels differ by >50-fold, there is no real need for a reference gene and normalization against total RNA is sufficient. If target levels differ by between six- and tenfold, a single reference gene may suffice, especially if comparisons are carried out between single cell lines. If two samples differ by <5-fold in their RNA levels, it is essential to use multiple reference genes, an approach that is robust and allows accurate normalization if fine measurements are to be made. Even then, the resolution of the particular assay remains dependent on the sample and variability of the chosen reference genes. For example, it will be far more difficult to find a set of reference genes that vary by <3-fold when using colorectal cancer samples from individual patients than if using colorectal cancer cell lines.

4   Considerations for the Future

The breakthrough of MIQE and its acceptance by the wider research community is welcome and the increasing inclusion of the various quality control parameters will undoubtedly result in the publication of peer-reviewed publications of a higher technical standard. It is unfortunate that it has taken such a long time to draft, publish, and disseminate these guidelines, as the peer-reviewed scientific literature now comprises thousands of publications that have erroneous conclusions based on inappropriate qPCR results. Anyone who has ever looked for publications supporting one or the other of two opposite viewpoints is very likely to find a number of publications supporting either position [23]. This is vexing and very likely will continue to lead to intellectual diversion as well as further investment in wasted cost and time.

However, another specter is even more thought-provoking and constitutes a logical extension of previously published information. The excellent investigations into the properties of RTs discussed above [14, 15] concluded that the type of RT, the priming strategy, and the amount of RNA used can generate significantly different results. It has been argued that this does not affect the ability to obtain comparable RT-qPCR, since the RT reaction is highly reproducible as long as the same experimental protocol and reaction conditions are used [24]. However, the RT step is not necessarily linear across different targets, with significant differences between the detection limits of different RTs that may be due to some components in the RT system that bias the subsequent PCR amplification [25]. Hence this viewpoint misses the inescapable truth that if four laboratories use four different RTs, four different priming strategies (gene-specific primers, hexamers, pentadecamers, oligo-dT), and different experimental protocols (widely different amounts of RNA, different volumes, different temperatures and times), they can end up with significantly different results despite following best practice protocols. The fact that each of these groups complies with the MIQE guidelines when publishing their data helps understand why the results may be different, but does not indicate which of the results is likely to be the correct one. These same authors concede, that we have no idea how the isolation yield varies among different mRNAs as differences in length, folding, localization in the cell, and complex formation with proteins are just some factors that may affect RNA extraction yield. This adds another level of error, and again it is perfectly conceivable that all extraction protocols are performed to the highest possible standards, that the RNA is quality assessed and handled appropriately, but that the additional variability introduced by the variable protocols will add to the RT-based errors, making any meaningful comparison of data very difficult. If this is then extended to the evaluation of reference genes, it is not inconceivable that different research groups will end up with different reference genes, and that the normalization of the RT-qPCR data will introduce yet one more level of variability. Importantly, all the results are technically accurate and deliver results that are repeatable and may even be reproducible, but depend entirely on how the experiments were performed. This is a real Achilles heel of the RT-qPCR technology, and is a problem that has not yet been addressed in a satisfactory manner. Perhaps the introduction of digital PCR will help somewhat reduce the distortion introduced by relative quantification, but by itself it does not tackle the problems of the RT step. Perhaps the solution is to focus our efforts on the quantification of proteins using PCR with proximity ligation or extension assays, and determine the levels of proteins, rather than the highly variable and, possibly, biologically less relevant levels of mRNA or even miRNAs. Proteins are, after all, the molecules that have function and identifying changes in the levels of proteins may be more informative than simply counting mRNA levels. Of course, protein-targeting introduces a whole new range of problems, but this should not distract from posing the question whether the only point of quantifying mRNA or miRNA levels by qPCR is to investigate a narrow set of regulatory mechanism, but that this approach is unlikely to yield information on the wider question of how and why cells alter their behavior in response to stimuli, or why normal cells develop into cancer cells. Changes to RNA transcript levels are, after all, a very small part of the overall mechanism of gene expression that involves huge numbers of proteins and their isoforms with different extents of posttranslational modifications, some cleaved and others complexed into active forms and all directly relevant to cell behavior (Fig. 5). On the other hand, many RNAs are useful as

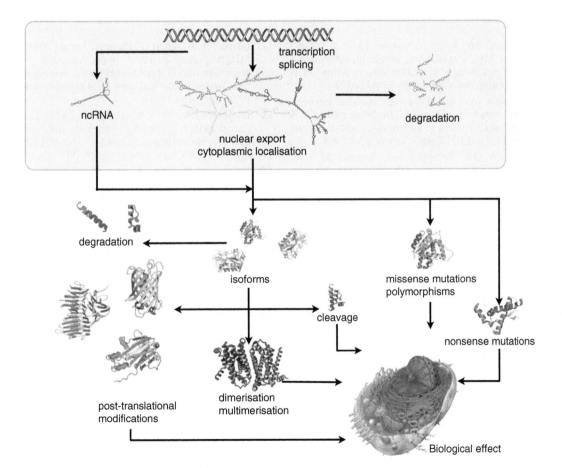

Fig. 5 The complexity of the gene expression pathway

diagnostic or prognostic biomarkers, in which case the use of RT-qPCR will continue to be necessary and valuable, but will have to be carried out with regard to the problems outlined above.

In conclusion, biologically relevant qPCR results depend on many parameters that must all come together in an appropriate, controlled, and transparent manner. These are highlighted by the MIQE guidelines, which are increasingly seen as a valuable sentinel safeguarding the reliability and reproducibility of qPCR-based conclusions. The urgent need for their implementation is vividly demonstrated by a recent publication that reviewed over 2,000 publications for compliance with the MIQE guidelines and found that the vast majority report inadequate or inappropriate information [26]. MIQE aims to make nucleic acid analysis not just easy but reliable and allow qPCR to exploit its wide range of applications that today range from the quantification of RNA to epigenetics and protein detection, using variations on essentially the same theme. Addressing one set of challenges opens up new ones, and as applications of this simple technique continue to diversify, there will be a greater need for more bespoke considerations.

References

1. Bustin SA, Benes V, Garson JA et al (2009) The MIQE guidelines: minimum information for publication of quantitative real-time PCR experiments. Clin Chem 55:611–622

2. Bustin SA (2013) Why there is no link between measles virus and autism. In: Fitzgerald M (ed) Recent advances in autism spectrum disorders, vol I, Intech. Rijeka, Croatia, pp 81–98

3. Kakisi OK, Robinson MJ, Tettmar KI et al (2013) The rise and fall of XMRV. Transfus Med 23:142–151

4. Roehr B (2012) Researchers find no link between XMRV and chronic fatigue syndrome. BMJ 345:e6331

5. Bohlenius H, Eriksson S, Parcy F et al (2007) Retraction. Science 316:367

6. Bustin SA (2010) Why the need for qPCR publication guidelines?—The case for MIQE. Methods 50:217–226

7. Abdel Nour AM, Pfaffl M (2011) Apple store, apple https://itunes.apple.com/app/miqeqpcr/id423650002?mt=8 or https://play.google.com/store/apps/details?id=com.biorad.miqeqPCR

8. Anonymous (2013) Announcement: reducing our irreproducibility. Nature 496:398

9. Shaffer C (2010) Routine lab method's accuracy called into question. Nat Med 16:349

10. Marx V (2013) PCR: living life amplified and standardized. Nat Methods 10:391–395

11. Bustin SA, Beaulieu JF, Huggett J et al (2010) MIQE precis: practical implementation of minimum standard guidelines for fluorescence-based quantitative real-time PCR experiments. BMC Mol Biol 11:74

12. Huggett JF, Foy CA, Benes V et al (2013) The digital MIQE guidelines: minimum information for publication of quantitative digital PCR experiments. Clin Chem 59:892–902

13. Bustin SA, Benes V, Garson JA et al (2011) Primer sequence disclosure: a clarification of the MIQE guidelines. Clin Chem 57:919–921

14. Stahlberg A, Hakansson J, Xian X et al (2004) Properties of the reverse transcription reaction in mRNA quantification. Clin Chem 50: 509–515

15. Stahlberg A, Kubista M, Pfaffl M (2004) Comparison of reverse transcriptases in gene expression analysis. Clin Chem 50: 1678–1680

16. Nolan T, Hands RE, Ogunkolade BW et al (2006) SPUD: a qPCR assay for the detection of inhibitors in nucleic acid preparations. Anal Biochem 351:308–310

17. Huggett JF, Novak T, Garson JA et al (2008) Differential susceptibility of PCR reactions to inhibitors: an important and unrecognised phenomenon. BMC Res Notes 1:70

18. Bustin SA (2000) Absolute quantification of mRNA using real-time reverse transcription polymerase chain reaction assays. J Mol Endocrinol 25:169–193

19. Huggett J, Bustin SA (2011) Standardisation and reporting for nucleic acid quantification. Accredit Qual Assur 16:399–405

20. Huggett J, Dheda K, Bustin S et al (2005) Real-time RT-PCR normalisation; strategies and considerations. Genes Immun 6:279–284

21. Tricarico C, Pinzani P, Bianchi S et al (2002) Quantitative real-time reverse transcription polymerase chain reaction: normalization to rRNA or single housekeeping genes is inappropriate for human tissue biopsies. Anal Biochem 309:293–300

22. Dheda K, Huggett JF, Chang JS et al (2005) The implications of using an inappropriate reference gene for real-time reverse transcription PCR data normalization. Anal Biochem 344: 141–143

23. Bustin SA, Mueller R (2005) Real-time reverse transcription PCR (qRT-PCR) and its potential use in clinical diagnosis. Clin Sci (Lond) 109: 365–379

24. Kubista M, Andrade JM, Bengtsson M et al (2006) The real-time polymerase chain reaction. Mol Aspects Med 27:95–125

25. Levesque-Sergerie JP, Duquette M, Thibault C et al (2007) Detection limits of several commercial reverse transcriptase enzymes: impact on the low- and high-abundance transcript levels assessed by quantitative RT-PCR. BMC Mol Biol 8:93

26. Bustin SA, Benes V, Garson J et al (2013) The need for transparency and good practices in the qPCR literature. Nat Methods 10: 1063–1067

Chapter 3

Selection of Reliable Reference Genes for RT-qPCR Analysis

Jan Hellemans and Jo Vandesompele

Abstract

Reference genes have become the method of choice for normalization of qPCR data. It has been demonstrated in many studies that reference gene validation is essential to ensure accurate and reliable results. This chapter describes how a pilot study can be set up to identify the best set of reference genes to be used for normalization of qPCR data. The data from such a pilot study should be analyzed with dedicated algorithms such as geNorm to rank genes according to their stability—a measure for how well they are suited for normalization. geNorm also provides insights into the optimal number of reference genes and the overall quality of the selected set of reference genes. Importantly, these results are always in function of the sample type being studied. Guidelines are provided on the interpretation of the results from geNorm pilot studies as well as for the continued monitoring of reference gene quality in subsequent studies. For screening studies including a large, unbiased set of genes (e.g., complete miRNome) an alternative normalization method can be used: global mean normalization. This chapter also describes how the data from such studies can be used to identify reference genes for subsequent validation studies on smaller sets of selected genes.

Key words qPCR, Selection of reference genes, Human candidate reference genes, Candidate reference microRNAs, geNorm

1 Introduction

Reverse transcription quantitative PCR (RT-qPCR) is currently the most popular method for the comparison of gene expression levels among biological samples. The reasons for the popularity of RT-qPCR are its sensitivity, flexibility, low cost of reagents and instruments, and the apparent ease to perform a qPCR experiment. The ease to generate qPCR data is in sharp contrast with the challenges to guarantee that the obtained results are reliable. One of the biggest challenges is proper normalization of the data.

Many different methods have been proposed to normalize qPCR data. An overview of basic normalization concepts is nicely reviewed in Huggett et al. [1]. Of all the options to normalize qPCR data, the use of reference genes (historically referred to as housekeeping genes) is undoubtedly the most popular approach. It is appealing not only for its practical simplicity but also for its

Roberto Biassoni and Alessandro Raso (eds.), *Quantitative Real-Time PCR: Methods and Protocols*, Methods in Molecular Biology, vol. 1160, DOI 10.1007/978-1-4939-0733-5_3, © Springer Science+Business Media New York 2014

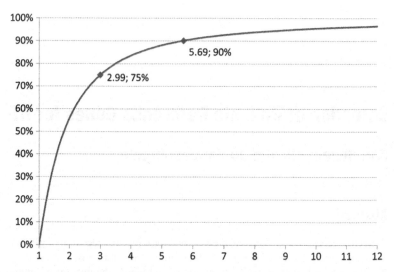

Fig. 1 Cumulative distribution of the normalization bias caused by using a single nonvalidated reference gene. Data from Vandesompele et al. [2]

potential to remove most of technical variation in cDNA concentrations between samples. However, for reference genes to meet this potential, they have to be stably expressed. Studies performed in 2002 by Vandesompele et al. and many others since then highlight the risk of blindly relying on the assumption of stable expression of reference genes [2]. The average difference in expression level of a gene of interest after normalization with any of two randomly selected nonvalidated reference genes is at least threefold in 25 % of the cases and even more than sixfold in 10 % of the cases (Fig. 1). In view of the modest degree of differential expression observed in many experiments, this bias is clearly unacceptable.

These results clearly indicate the need for proof that selected reference genes are stably expressed.

2 geNorm

A naive approach to verify stable expression is the comparison of raw *Cq* values for reference genes between different samples, looking for genes with minimal variation in *Cq* values. However, this approach is flawed because it assumes equal cDNA concentrations for all samples. Of note, cDNA inputs are often standardized by means of equal total RNA inputs. As total RNA consists mostly of ribosomal RNA and as mRNA:rRNA ratios are quite variable depending on cell type and condition, equal cDNA input amount is an unreliable measure to normalize or to validate reference genes. More robust methods that are independent of cDNA input concentrations are required to quantify the degree of variation in expression levels for reference genes. In 2002, Vandesompele et al. described the geNorm approach to identify stably expressed reference genes. This method is robust and independent of cDNA input amount (as long as the same amount is used to measure all candidate reference genes in that sample). Since then, a number of alternative methods have been published (for overview see book chapter on "Reference gene validation software for improved normalization" [3]). The choice between any of these algorithms is less important than the fact of actually using a valid method to determine the stability of reference genes. In most cases, results will be equivalent when using different approaches.

An experiment to identify stably expressed reference genes prior to conducting the real experiment (in which the gene(s) of interest are measured) is referred to as a geNorm pilot study. Such a study typically consists of eight or more candidate reference genes evaluated for ten or more representative samples. Candidate reference genes may be selected from a list of usual suspects (examples for human (Table 1)), from commercial reference gene kits

Table 1
Human candidate reference genes

Official symbol	Official name	Ensembl gene ID	Entrez gene ID	Refseq IDs
ACTB	Actin, beta	ENSG00000075624	60	NM_001101
B2M	Beta-2-microglobulin	ENSG00000166710	567	NM_004048
GAPDH	Glyceraldehyde-3-phosphate dehydrogenase	ENSG00000111640	2597	NM_002046, NM_001256799
GUSB	Glucuronidase, beta	ENSG00000169919	2990	NM_000181
HMBS	Hydroxymethylbilane synthase	ENSG00000256269	3145	NM_001258209, NM_000190, NM_001258208, NM_001024382
HPRT1	Hypoxanthine phosphoribosyltransferase 1	ENSG00000165704	3251	NM_000194
PPIA	Peptidylprolyl isomerase A (cyclophilin A)	ENSG00000196262	5478	NM_021130
RPL13A	Ribosomal protein L13a	ENSG00000142541	23521	NM_012423, NM_001270491, NR_073024
RPS18	Ribosomal protein S18	ENSG00000231500	6222	NM_022551
SDHA	Succinate dehydrogenase complex, subunit A, flavoprotein (Fp)	ENSG00000073578	6389	NM_004168
TBP	TATA box binding protein	ENSG00000112592	6908	NM_003194, NM_001172085
TUBB	Tubulin, beta	ENSG00000196230	203068	NM_178014
UBC	Ubiquitin C	ENSG00000150991	7316	NM_021009
YWHAZ	Tyrosine 3-monooxygenase/tryptophan 5-monooxygenase activation protein, zeta polypeptide	ENSG00000164924	7534	NM_003406, NM_001135699, NM_145690, NM_001135700, NM_001135701, NM_001135702

(e.g., from PrimerDesign), from the literature, from experimental data in the lab (e.g., from microarray or RNA-seq), or simply from assays available in the lab. It is important to avoid selecting multiple genes from the same pathway or functional class because strong coregulation may interfere with proper analysis of expression stability. The number of candidate reference genes to be tested varies depending on sample heterogeneity and required experimental accuracy and precision. More challenging experiments with complex, heterogeneous samples and small differences in expression of the gene of interest will require a larger set of candidates to allow selection of suitable reference genes. The selection of samples needs to be representative of the actual study to be performed since different sample types may result in a different selection of reference genes. If sample subgroups exist (e.g., treated and untreated), they need to be equally represented in the geNorm pilot study to avoid selecting genes that are stably expressed in one group, but not across all samples.

Three types of information can be extracted from a geNorm analysis:

1. Ranking of candidate reference genes according to their stability.

2. The optimal number of reference genes for the analyzed sample type.

3. An assessment of the overall stability of selected reference genes (Fig. 2).

Candidate reference genes are ranked according to their M-value, a measure for relative expression stability. Unstable reference genes with high M-values are sorted to the left; the best reference genes are found on the right side. The selection of most stably expressed genes will vary from sample type to sample type. Also the optimal number of reference genes is variable and can be determined based on the V-value. The V-value is an indication for how much difference it makes when using an extra reference gene for normalization. If the added value of adding one more is limited (guideline: V below 0.15), one could as well leave it out. The average M-value for the optimal number of reference genes is an indication for their combined quality to normalize your data. Based on many sets of empirical data we suggest the following guidelines. Average M-values below 0.2 are typically seen when evaluating reference targets using genomic DNA as input (e.g., for CNV analysis), or when reference targets are very stably expressed. Average M-values between 0.2 and 0.5 are typically seen when evaluating candidate reference targets on a homogeneous set of samples (e.g., untreated cultured cells, or blood from normal individuals). Average M-values in the 0.5–1 range are expected when evaluating candidate reference targets on a heterogeneous set of samples

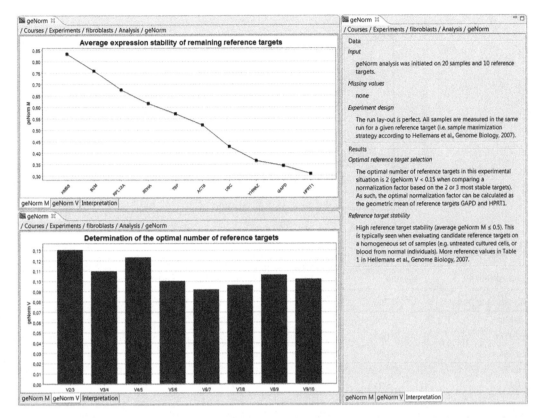

Fig. 2 Example results for a geNorm analysis on fibroblasts, generated by the qPCR data analysis software qbase+

(e.g., treated cultured cells, cancer biopsies, or samples from different tissues). Average *M*-values larger than 1 indicate low reference target stability. In those cases, it might prove useful to evaluate other candidate reference genes.

Reference genes selected in a geNorm pilot can be used in all subsequent experiments analyzing the same sample type in similar conditions. It is advisable to continue monitoring the stability of selected reference genes. This quality control is only possible if two or more reference genes are included in an experiment. Reference gene stability can be quantified by calculating geNorm *M*-values, or alternatively by inspecting the variation (expressed as CV values) on the normalized relative quantities of tested reference genes. This test is very valuable to exclude the possibility that one of your treatments or manipulations unexpectedly results in altered expression levels for one of your reference genes, and it comes at no extra cost.

Performing a geNorm pilot study does involve extra work to acquire reference gene assays, to run the pilot reactions, and to analyze and interpret the results. Because of this, many people take a pragmatic approach to the issue of reference gene selection. By consulting their colleagues, peers, or the literature they make an

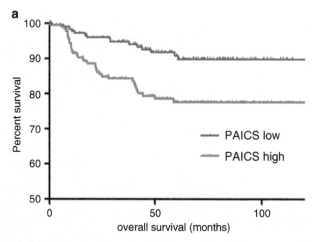

b

normalization	Cox regression p-value	significant	fold change in p-value	Cox regression hazard ratio	change in hazard ratio
5 reference genes	0.002	yes		3.990	
AluSq	0.005	yes	2.5	2.292	-43%
HPRT1	0.006	yes	3.0	2.381	-40%
SDHA	0.085	no	42.5	1.633	-59%
HMBS	0.186	no	93.0	1.434	-64%
UBC	0.232	no	116.0	1.331	-67%

Fig. 3 (**a**) Kaplan–Meier survival curve for children with neuroblastoma. 366 patients were dichotomized with respect to the median PAICS mRNA expression. Normalization was done with five stably expressed reference genes (UBC, HPRT1, SDHA, HMBS, and AluSq). The survival rates of patients with PAICS expression higher than the median were significantly worse than those with PAICS expression lower than the median ($p = 0.002$) (data from Vermeulen et al. [5]). (**b**) Table showing increased p-value of the Cox proportional hazards test on data from the same experiment normalized with only a single reference gene

educated guess on suitable reference genes. For easy and common sample types this may be a valid approach, as long as the necessary follow-up quality controls, as described above, are included in all experiments. For more difficult samples or demanding studies it is still advisable to perform a full-blown geNorm study to identify the optimal set of reference genes. The benefits of doing so have been demonstrated over and over again. A good set of reference genes enables the removal of most of the technical (experimentally induced) variation. This leads to the possibility to detect smaller differences in expression levels, down to 50 % in real patient samples [4], and to more significant results [5]. The example in Fig. 3 shows how a p-value of 0.002 can be obtained by the combined use of five validated reference genes (of note, clinical cancer samples often display large expression heterogeneity, requiring more reference genes than usual) (Table 1). None of the individual reference genes are as good as the combination (3 out of 5 even give rise to a nonsignificant result). Any randomly selected, nonvalidated reference gene is likely to perform much worse. PAICS is only one of the 59 tested genes in Vermeulen et al. When doing the same analyses for the other genes, single reference gene normalization results in loss of significance in 7 % of the cases, with an overall increase in p-value of 60-fold (range 2–121,308). In addition, the hazard ratio (risk of death) changed for all tested genes, with an average impact on the magnitude by 45 %. The above results clearly indicate that multi-gene normalization allows to remove the noise in the data and to focus on the true biologically meaningful and clinically relevant differences.

3   Reference for microRNAs

When it comes to microRNA data normalization, there are no predefined sets of candidate reference microRNAs. Historically, qPCR users relied on small nuclear or small nucleolar RNAs for normalization (e.g., U6 or RNU44). But, similar to our arguments against ribosomal RNA for mRNA normalization, we advise against the use of these internal controls. Both rRNA and sn(o)RNA are transcribed from a different RNA polymerase and have different functions than mRNA and miRNA, respectively. Even without these arguments, it's never wise to simply rely on one or few (popular) reference genes; you really should test if the candidate reference genes are stably expressed in your experimental condition.

In 2009, we published a new method for even more accurate normalization when a large unbiased set of genes is measured [6]. We applied the method for normalization of microRNA expression profiling studies in which typically a few hundred microRNAs are measured. The method has since then been perfected [7] by attributing equal weight to each individual miRNA during normalization. The improved method is implemented in the qbase + software as "global mean normalization method".

While the above referenced method works great, it requires many microRNAs to be measured. For follow-up studies, one typically is only interested in the validation of (part of) the differentially expressed microRNAs. To normalize that type of data, we recommend the use of multiple stably expressed microRNAs. We propose the following procedure to find such stably expressed candidate reference microRNAs:

1. Convert Cq values in non-normalized relative quantities (RQ = $2^{\Delta Cq}$, with ΔCq being the difference in Cq between a reference sample (calibrator) and the sample of interest).

2. Normalize relative quantities using the global mean method.

3. Log transform the normalized data.

4. Calculate the standard deviation per microRNA across all tested samples.

5. Select candidate microRNAs that (a) have data for all samples, (b) have lowest standard deviation, (c) do not belong to the same miR family (use only the best miR per family; miR families can be inspected in a special miRBase file available on ftp://mirbase.org/pub/mirbase/CURRENT/miFam.dat.gz). We recommend selecting at least three (five or more is better) candidate reference microRNAs for use in the final experiment.

6. Verify in your final experiment that these candidate microRNAs are stably expressed (low M-values, guidance is offered in the geNorm report in qbase + software).

If you do not have access to large-scale microRNA profiles, consider profiling a few representative samples after which you can follow the procedure outlined higher. If that is not an option, you should setup a classic geNorm pilot experiment with sn(o)RNAs and published candidate reference microRNAs (ideally in the same type of samples). Typically, eight candidate small RNAs are measured in at least ten representative samples. The geNorm report in qbase + will help you to identify the most stably expressed genes and will suggest how many genes to use to achieve optimal normalization.

References

1. Huggett J, Dheda K, Bustin S et al (2005) Real-time RT-PCR normalisation; strategies and considerations. Genes Immun 6:279–284
2. Vandesompele J, De Preter K, Pattyn F et al (2002). Accurate normalization of real-time quantitative RT-PCR data by geometric averaging of multiple internal control genes. Genome Biol 3:RESEARCH0034
3. Vandesompele J, Kubista M, Pfaffl MW (2009) Real-time PCR: current technology and applications. In: Logan J, Edwards K, Saunders N (eds) Caister Academic, Norfolk, England, pp 1–27
4. Hellemans J, Preobrazhenska O, Willaert A et al (2004) Loss-of-function mutations in LEMD3 result in osteopoikilosis, Buschke-Ollendorff syndrome and melorheostosis. Nat Genet 36: 1213–1218
5. Vermeulen J, De Preter K, Naranjo A et al (2009) Predicting outcomes for children with neuroblastoma using a multigene-expression signature: a retrospective SIOPEN/COG/GPOH study. Lancet Oncol 10:663–671
6. Mestdagh P, Van Vlierberghe P, De Weer A et al (2009) A novel and universal method for microRNA RT-qPCR data normalization. Genome Biol 10:R64
7. Barbara D, Mestdagh P, Hellemans J et al (2012) miRNA expression profiling: from reference genes to global mean normalization. Methods Mol Biol 822:261–272

Chapter 4

Introduction to Digital PCR

Francisco Bizouarn

Abstract

Digital PCR (dPCR) is a molecular biology technique going through a renaissance. With the arrival of new instrumentation dPCR can now be performed as a routine molecular biology assay. This exciting new technique provides quantitative and detection capabilities that by far surpass other methods currently used. This chapter is an overview of some of the applications currently being performed using dPCR as well as the fundamental concepts and techniques this technology is based on.

Key words Digital PCR, Nucleic acid quantitation, Viral analysis, Copy number variations, Copy number alterations, Rare mutation detection, Rare mutation abundance, Gene expression analysis

1 Introduction

Digital PCR (dPCR) represents a third iteration of PCR that enables precise, highly sensitive quantification of nucleic acids. It combines classic PCR reaction kinetics with fluorescence-based detection strategies commonly used in real-time quantitative PCR (qPCR). dPCR involves partitioning a single PCR reaction into hundreds or thousands of subreactions under conditions where some of the subreactions amplify, indicating there is target nucleic acid present in that partition, and some of the subreactions do not, indicating that no target is present. The subreactions are individually analyzed for the amplification of interest. The ratio of these positive subreactions to negative subreactions can be used to accurately determine the initial number of target molecules within the original sample. Due to its digital nature, dPCR provides direct, highly sensitive and absolute nucleic acid quantification without the use of an external reference (standard) curve and is less dependent on PCR reaction amplification efficiency than qPCR reactions (Figs. 1 and 2).

1.1 History

dPCR was first presented by Sykes et al. in 1992 [1] who described serially diluting a PCR template (threefold) down to a concentration, where within a replicate set (ten replicates), some of the

Roberto Biassoni and Alessandro Raso (eds.), *Quantitative Real-Time PCR: Methods and Protocols*, Methods in Molecular Biology, vol. 1160, DOI 10.1007/978-1-4939-0733-5_4, © Springer Science+Business Media New York 2014

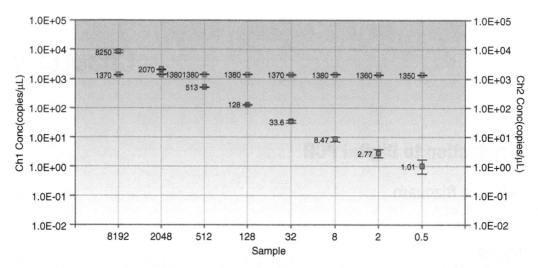

Fig. 1 Single reaction 2plex ddPCR reaction results. Fourfold dilution series of *S. aureus* DNA (*blue boxes*) in the presence of a constant background of Human genomic DNA (*green boxes*). Each target was probed for using a different colored probe (FAM for *S. aureus* and Hex for Human). *Y* axis represents ddPCR concentration; *X* axis represents estimated concentration using classical quantitation methods. Error bars represent 95 % confidence intervals of combined partitioning error and sample distribution error. Data generated on the QX100 Droplet Digital PCR System

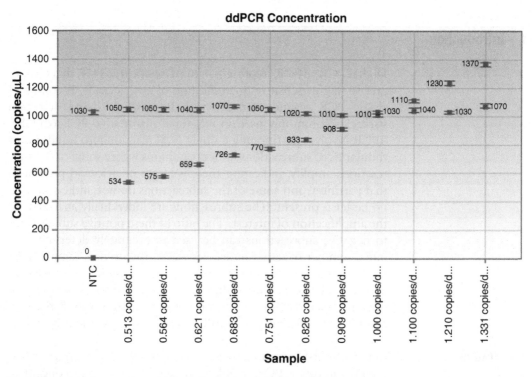

Fig. 2 Merged duplex reactions. 10 % dilution series of *S. aureus* DNA (*blue boxes*) in the presence of a constant background of Human genomic DNA (*green boxes*). Combination of 4 × 20,000 partition reactions per sample (80,000 partitions per duplex sample). Each target was probed for using a different colored probe (FAM for *S. aureus* and Hex for Human). *X* axis represents estimated concentration using classical quantitation methods. Error bars represent 95 % confidence intervals of combined partitioning error and sample distribution error

replicate PCR reactions amplified and some did not. These results were used in a Poisson analysis calculation to determine initial copy number. At the time, due to the reaction volumes used, large number of wells required to run the serial replicates and the downstream analysis required, this approach was deemed impractical for widespread use.

The term Digital PCR was first used by Vogelstein and Kinzler in 1999 [2] to describe a rare mutation abundance experiment for the oncogene kRAS where larger sets of replicate PCR reactions were probed with Molecular Beacons and analyzed for red/green fluorescent signals to quantify wild-type and mutant molecules. These experiments were run in 96 well plates using 7 μl reactions, but there were sights on increasing the number of replicates to 384 and 1,536 as these plate formats became available. The analysis of positive wells providing a count of one and negative wells contributing a count of zero, subsequently used for Poisson analysis, led to the term Digital PCR.

With the arrival of new technologies and commercially available instruments, dPCR is going through a renaissance; becoming a practical tool for nucleic acid analysis where other current methods lack the resolution, sensitivity, and cost points required. Applications such as absolute quantification, high resolution gene expression, miRNA analysis, copy number variations, and rare mutation abundance that are currently difficult and tedious using qPCR and other techniques can be analyzed quickly and effectively using dPCR.

2 Applications

There are many applications that can benefit from the higher resolution and partitioning effects that are conferred by dPCR. Some that jump to the forefront are described in the following sections.

2.1 Viral/Pathogen Detection and Quantitation [3–5]

Currently the quantitation of viruses and pathogens requires highly defined standards, and assays must be performed under identical conditions and at the same time as these. A common obstacle is the sometimes low level of target molecules and the presence of minor inhibitors in the sample that skew the accuracy of the quantitative results. dPCR does not require standards and tolerates the some inhibition.

2.2 High Resolution Gene Expression Analysis

Determining differences in expression levels between samples at levels less than twofold can be a challenge using technologies such as qPCR [6, 7]. A Cq on its own has little value unless compared to another sample, adjusted for amplification efficiency and normalized to reference genes (each of which is dependent on other samples and on assay efficiency). The propagation of small errors can generate large error bars that make small difference analysis difficult if not impossible. dPCR generates a higher resolution absolute numerical output independent of other samples that when normalized to reference genes generally yields more statistically significant results.

2.3 Copy Number Variation [8–11]

As with gene expression analysis, dependence on other samples for significance (in gels, qPCR, etc.) makes accurate copy number determination difficult. These can be run in duplex reactions in dPCR providing numerical absolute results that have a larger integer dynamic range.

2.4 Single Cell Analysis

Single cell samples are often preamplified prior to any downstream analysis. The greater the number of amplification cycles, the greater the bias in favor of some genes to the detriment of others. dPCR can yield more accurate results at lower target levels, thus requiring less preamplification.

2.5 Rare Mutation Abundance Detection [12, 13]

Finding a somatic mutation in a disease containing sample (tissue, blood, etc.) can be difficult due to their low abundance in a large background of normal sample. The partitioning and enrichment effects present in dPCR allow for the detection and quantitation of these proverbial needles in haystacks.

2.6 Controls Quantitation

Proper controls are essential in the fields of clinical diagnostics, food testing, and instrumentation installations and qualifications. Determining the quality of nucleic acid controls is difficult, as most quantification methods can be either inaccurate, as is the case with using optic density, or dependent on other measurements, as is the case with qPCR. dPCR provides standalone absolute quantitative results and is becoming popular with companies and organizations that provide specific controls.

2.7 Library Quantitation for Next Gen Sequencing

Finding the correct amount of library to load on an NGS platform is not arbitrary. Underloading results in low read levels while overloading results in large swaths of unusable data. Proper quantification of library yields optimal results. dPCR generates a numerical output regardless of varying amplicon lengths which can bias quantitative analysis using other methods.

2.8 GMO and Food Testing [14]

Many food stocks require molecular testing to ascertain their content and safety for human consumption. It is common to encounter the presence of inhibitory molecules that are costly and/or difficult to remove from the sample and can cause misrepresentation of the actual content. dPCR is less sensitive to minor inhibition and maintains quantitative integrity without the need of an external standard reference sample set.

3   Current Practice

The process of digital PCR is simple; a PCR reaction is prepared with the usual components (primers, dNTPs, polymerase, buffers, ions, etc.) and a reporter molecule (dye or probe). Once thoroughly mixed, the reaction (typically 20 µl) is partitioned into subreactions that are physically separated from one another. This partitioning can be done using a number of ways; using chambered silica wafers on a microfluidic chip [15–17] using microarrays [18], spinning microfluidic discs [19], and droplet techniques based on oil–water emulsions [12]. There are currently on the market various instruments that allow the samples to easily be partitioned into hundreds, thousands, or millions of subreactions providing varying degrees of resolution, precision, and cost per sample tested (Fig. 3).

Once partitioned, the reactions are then amplified using standard PCR cycling conditions. Post amplification, the subreactions are probed for the presence or absence of amplification and tallied. Target molecule amplification is typically determined using either a hydrolysis probe or through the use of a fluorescent DNA-binding dye (Figs. 4 and 5).

dPCR takes advantage of being an end point analysis of each subreaction. Amplification efficiency variations from sample to sample due to minor inhibitors or delayed amplification start (target accessibility) have substantially less of an impact on quantitative results than when using other techniques. General inhibition, as long as it does not halt the reaction completely, will allow analysis to move forward as long as positive events can be clearly differentiated from negative ones. Under normal circumstances, individual positive events that do not have final fluorescent amplitude similar to that of the rest may suffer from delayed amplification caused by primer accessibility issues (such as poor denaturing or a point mutation on the primer annealing site of the specific DNA molecule) or by a point mutation on the probe annealing site. In all of

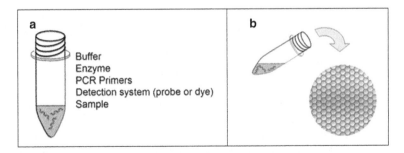

Fig. 3 Components of a Digital PCR reaction and Partitioning. dPCR reactions contain similar components to that typically used in a qPCR reaction. These components should be thoroughly mixed and then partitioned into uniform subpartitions

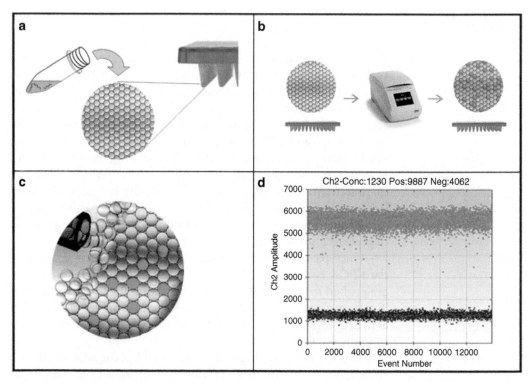

Fig. 4 (**a**) Partitioned reactions are placed in a vessel compatible for the PCR amplification to take place. (**b**) Amplification can take place in a standard thermocycler using typical qPCR protocols. For example 95 °C for 10 min followed by 40 cycles of 95 °C for 30 s and 60 °C for 1 min. No optical analysis is required at the amplification phase. (**c**) Individual subreactions are collected and subsequently interrogated for the presence of a positive fluorescent signal. (**d**) Individual reactions are deemed either a positive or negative event, counted, and the concentration is calculated

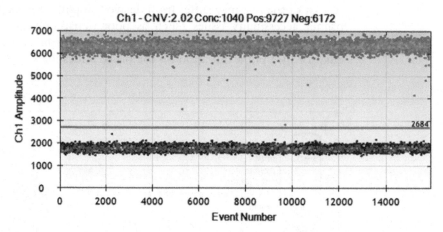

Fig. 5 Temporal plot of a dPCR reaction. *X* axis represents individual subreactions in the order they were analyzed. *Y* axis represents fluorescent signal generated in each corresponding subreaction. *Blue dots* are deemed as positive events; *grey dots* are deemed negative. *Purple line* is the threshold for determining what is positive from negative (can be set automatically by software algorithms or manually as in the case above)

these cases, since the target DNA molecule is present, the events should be considered positive.

At first view, many would simply think that the total target molecule count in our reaction is simply the number of positive subcompartments. This is somewhat but not entirely correct. If target molecules are present in low amounts (a few percent of the amount of subpartitions used), then random distribution would probably do a reasonable job at distributing them somewhat evenly. When target molecule numbers are above a few percent versus the number of partitions, then there is an ever increasing probability that multiple targets will co-migrate to the same compartment. This is when Poisson distribution analysis comes into play.

The positive and negative subreaction counts are used to determine the total target molecule count. The ratio of positives to total (p) is used to determine the average number of targets per subreaction (λ) using the Poisson formula below and then multiplied by the number of subreactions per µl. These calculations are typically generated in the accompanying analysis software packages that are provided with dPCR technologies. As reaction volumes may vary depending on the type of experiment and instrument used, results are generally presented as copies per µl of reaction (Fig. 6).

Of particular interest to many researchers is the absence of a standard reference curve. In qPCR, regardless of the type of assay run, all data points (Cq's, Ct's, etc.) are meaningless on their own. These data points are dependent on others (ref curve, other samples, etc.) for their significance. In dPCR, each reaction is a stand-alone and absolute result. Quantitation is determined by the attributes of each reaction on a positive versus negative basis of the sub partitions, independent of assay efficiency or the day to day variables of reaction mix preparation, minor reaction inhibition, and user inconsistencies. This is extremely useful and practical for samples that are either analyzed at different time points, have low levels of target molecules, and for samples for which generating an accurate absolute standard curve is difficult or impractical.

The practice of performing a dPCR reaction has become simple, practical, and affordable. The key to successful dPCR

$$\lambda = - \ln(1-p)$$

For example 20% of partitions are positive in a 20 000 partition experiment where each partition is 1nl in volume.

Average copy number per compartment   = - ln(1-p) = - ln(1-.2) = (.223)

Copy number per microliters = 0.223 x 1000 = 223

Copy number per 20 microliters reaction   = 223 x 20 = 4462

Fig. 6 Typical calculations used to determine target copy number

experiments lies in proper sample mixing and partitioning. dPCR calculations are based on Poisson analysis and as such the principles of random distribution must be considered. Proper reaction mixing allows for all the reaction components to be evenly distributed within the subcompartments whereas improper mixing will generate compartments with poor sample distribution and varying reaction mix concentrations. Partition size uniformity is also important as compartments of uneven sizes skew the statistical distribution. Finally, the number of partitions used plays a role in the quantitative resolution; the larger the number of partitions, the greater the resolution (within reasonable limits).

4   Calculations to Consider

One of the most common misconceptions about dPCR is that the sample must be diluted down to a level such that one and only one target molecule is present in each subreaction. Random distribution, also known as Poisson distribution when it describes small numbers of events, dictates that the template DNA will not be evenly distributed amongst all the subreactions. Randomly some partitions will have more than a single target molecule co-migrate within it. For limiting dilutions to work (without Poisson analysis) and have minimal effect on quantitation values would require approximately 100-fold excess partitions versus target molecules. This is not only impractical and costly, but still carries a certain error with it. In practical real world experiments, the number of target molecules analyzed can greatly exceed the number of partitions used (Fig. 7).

When looking at Poisson distribution patterns as they relate to dPCR, it is important to note that at least one subreaction partition must remain empty (or negative) of the target for the calculations to be possible. If all partitions are positive, it is impossible to determine the target copy count as one could not predict the distribution pattern (tens? hundreds? or thousands? of target molecules per partition). The number of partitions therefore dictates the dynamic range for quantitation of a given reaction. As the number of negative subpartitions is more variable as we get closer to the upper end of the dynamic range, it is recommended to use target concentrations in the first 60 % or so of the total dynamic range. As presented in Table 1, a 10,000 partition reaction should have a maximum target molecule amount of approximately 55,000. Experiments that require a larger dynamic range than can be provided by a single reaction can be performed by "pooling" reactions (i.e., combining five 20,000 partition reactions to equate to one 100,000 partition reaction) (Table 1).

Due to the digital and absolute nature of dPCR results, one can calculate error probability from a single reaction. Errors other

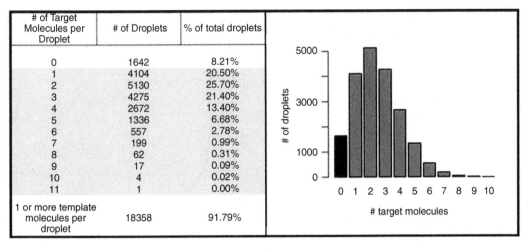

# of Target Molecules per Droplet	# of Droplets	% of total droplets
0	1642	8.21%
1	4104	20.50%
2	5130	25.70%
3	4275	21.40%
4	2672	13.40%
5	1336	6.68%
6	557	2.78%
7	199	0.99%
8	62	0.31%
9	17	0.09%
10	4	0.02%
11	1	0.00%
1 or more template molecules per droplet	18358	91.79%

Fig. 7 Example of expected Poisson distribution. 50,000 copies of target molecules in a 20 µl reaction are partitioned into 20,000 droplets (1 nl in size). Although the average number of copies per droplet is 2.5, Poisson distribution predicts how the distribution of target molecules within the droplets will occur for the entirety of the reaction

Table 1
Approximate dynamic range of a dPCR reaction as a function of number of subpartitions

Partitions	Approximate dynamic range	Practical dynamic range
16	1–44	1–26
64	1–266	1–160
256	1–1,406	1–875
1,000	1–6,907	1–4,144
5,000	1–42,585	1–25,551
10,000	1–92,103	1–55,261
20,000	1–198,069	1–118,841
50,000	1–540,988	1–324,592
100,000	1–1,151,292	1–690,775

than those generated upstream of the dPCR process (pipetting) are generally attributable to two sources: partitioning distribution error and subsampling errors. Partitioning errors are those associated with the random distribution of target molecules into the partitions in a predictable manner. For example, the distribution of targets in Fig. 7 may vary from one identical sample to the next due to random distributions of the target molecules such that the

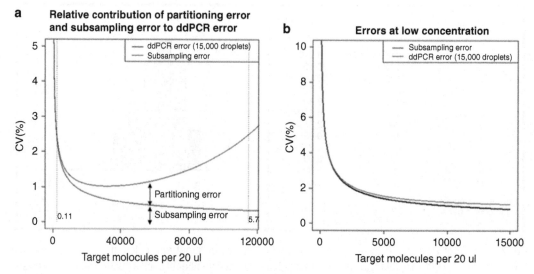

Fig. 8 Error contributions in Digital PCR. (**a**) Relative contributions to dPCR results due to partitioning and sub-sampling over the practical dynamic range. (**b**) Error contributions at lower concentrations

number of empty partitions may be 1,642 in one sample and 1,622 in the second. This will cause the Poison calculated value to differ somewhat between samples. Partitioning error will vary as a function of the total number of partitions used. The larger the number of partitions the smaller the error and vice versa. Generally, the use of more than 10,000 partitions per reaction confers a partitioning error that is much smaller than other more significant errors such as pipetting uniformity and subsampling error. Subsampling errors occur from the irreproducibility in the collection of a sample at low concentration from a larger pool. For example if we have 100 μl of total sample containing 100 target molecules and collect 10 μl samples for our quantitative analysis, Poisson distribution will again dictate that we may not always collect exactly 10 target molecules per sampling. We may collect 8, 9, 10, 11, 12, etc. molecules at a draw. This subsampling effect follows strict mathematical rules and there is no way around this error short of using whole samples (complete entity such as cell, organ, fluid in organism, etc…) and is often underreported in qPCR experiments. An example of subsampling error is presented in Fig. 8.

Because the Poisson equation used in dPCR uses the ratio of positive to total partitions, the total of number of partitions can vary from sample to sample yet still yield similar results. If one sample has 20 % positive subreactions calculated from a total of 20,000 partitions and a second sample has 20 % of its subreactions positive from a total of 18,000 partitions, both will yield the same result. What will vary is the 95 % confidence interval attributed to the result, with the sample with lower partition numbers having a slightly larger confidence interval.

In comparison with real-time quantitative PCR, dPCR results are absolute in nature regardless of the type of experiment being run (abs quant, gene expression, copy number variation, etc.). All subsequent calculations are performed using numbers (not in semi-log as when comparing Cq's). When analyzing copy number variation results, the target of interest amount is simply divided by that of the reference. When looking at gene expression, the target of interest is normalized to the geometric average of properly validated reference genes [20]. Error propagation is also calculated using standard methods.

5   Enrichment Effect

One of the principal advantages of dPCR is a synthetic enrichment effect of lower abundance target molecules versus higher abundance target molecules that occur in multiplex reactions.

Bulk PCR generally reactions work reasonably well amplifying and quantitating a single nucleic acid target, but require additional optimization for duplex and multiplex reactions where varying levels of target molecules are present. A common occurrence in duplex PCR reactions is the monopolization and depletion of reaction elements (polymerases, dNTPs and reaction components) by the amplification of the high abundance target molecules to the detriment of the lower abundance target the starves out and fails to amplify properly (Fig. 9).

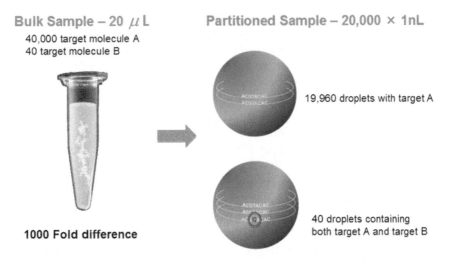

Bulk Sample – 20 μL

40,000 target molecule A
40 target molecule B

1000 Fold difference

Partitioned Sample – 20,000 × 1nL

19,960 droplets with target A

40 droplets containing
both target A and target B

Fig. 9 Enrichment effect of low target molecules. 40,000 molecules of target A will average 2 target molecules per PCR subcompartment when partitioned into 20,000. This makes the detection and quantitation of lower abundant target B much simpler

Partitioning the reaction has a diluting effect on the more abundant target, thus bringing it closer to the lower abundant one. In the example presented in Fig. 9, the bulk reaction has a 1,000-fold difference in the amount of target A versus target B and under multiplexing conditions where we are trying to quantitate both in a single reaction, the lower abundance target may fall out and be either underrepresented or fail to amplify. Although partitioning will not distribute target A evenly into 20,000 compartments of 2 target molecules (Poisson distribution strikes again) we can expect that their number will vary between 0 and 10 per subcompartment (averaging 2). The 40 molecules of target B will most likely migrate into 40 subcompartments where they will be at worst, at one tenth the concentration of target A. At these ratios, they have a much better chance of amplifying and subsequently being detected and quantified.

This enrichment effect simplifies the multiplexing optimization process which although should still be performed in dPCR is much simpler than with qPCR.

6   Multiplexing

Multiplexing in dPCR involves amplifying and detecting two or more target amplicons within a single PCR reaction. This is typically performed using fluorescently labeled probe-based detection strategies similar to those used in qPCR, although in dPCR, it can also be performed using DNA-binding dye-based assays. Most dPCR platforms provide two channels for the detection of fluorescent molecules. This allows for the quantitation of two or more individual targets. The first channel is normally FAM, and the second is normally HEX or VIC. The choice of these fluors stems from their use in qPCR.

Multiplexing is simpler in dPCR than in qPCR, due to the partitioning and enrichment effect. Many off the shelf assays can be combined and give excellent results. Nonetheless, due diligence requires some validation of these assays and occasionally optimization. The goal is to have two independent reactions that coincidentally occur within the same subpartitions.

Basic multiplexing results look similar to singleplex results, with the simple addition of a second set of charts corresponding to the second dye layer. Here again there must be proper discrimination between positive and negative signals. Advanced analysis using a 2D scatter plot can provide information on the overall quality of the assays and their interaction (if any). Clustering analysis is generally provided by the accompanying software package, although as with single target analysis, clustering can be performed manually. The calculations used to determine initial copies of target molecules follow the same principles as those used in a singleplex reaction (Figs. 10 and 11).

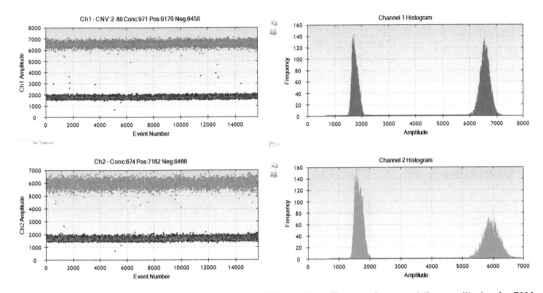

Fig. 10 Temporal and histogram plots of a duplex dPCR reaction. *Top panels* present the amplitude of a FAM label, *lower panel* HEX labeled reactions

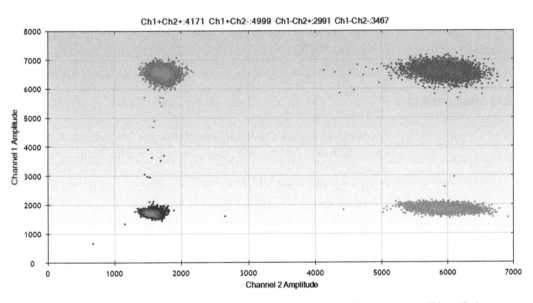

Fig. 11 2D amplitude graph of duplex dPCR reaction. *Grey dots* (*bottom left*) represent partitions that are negative for both targets. *Blue dots* (*upper left*) represent partitions that are positive for target probed using FAM labeled probe. *Green dots* (*bottom right*) represent partitions that are positive for the target probed using a HEX labeled probe. *Red dots* (*upper right*) represent partitions that are positive for the amplification of both targets

Multiplexing using DNA-binding dyes requires a single channel, but requires the amplicon products to be of different sizes. This will generate three populations of fluorescent products. Two form having a single target targets in each of the compartments and a third from the combination of both targets identification of the appropriate populations can be established though the use of singleplex controls.

Higher multiplexing levels can be achieved using a 2 color instrument by using varying concentrations of probes for different targets that will map to a different location on the 2D plot for different targets.

7   Conclusion

dPCR is a new iteration of the PCR process. It provides outstanding accuracy for the quantitation of nucleic acids and dramatically facilitates the detection and analysis of low abundance targets. Quantitative results produced are absolute and are not dependent on outer external reference samples. The practical and technical obstacles to the everyday use of dPCR have been overcome through the arrival of new instruments and reagents. dPCR will most likely become a routine technique in most molecular biology laboratories in life science research, in nucleic acid screening facilities, and in clinical diagnostics. Data quality, cost, and time to results make a valuable tool for quick and accurate detection and quantitation of DNA and RNA.

Acknowledgements

I would like to acknowledge and thank my colleagues at Bio-Rad Laboratories; Adam McCoy, George Karlin-Neumann, Jack Reagan, Svilen Tzonev, and all the members of the Digital Biology Center for their contributions to this overview and for their tireless efforts in the development and advancement of digital PCR.

References

1. Sykes PJ, Neoh SH, Brisco MJ et al (1992) Quantitation of targets for PCR by use of limiting dilution. Biotechniques 13:444–449
2. Vogelstein B, Kinzler KW (1999) Digital PCR. Proc Natl Acad Sci U S A 96:9236–9241
3. Strain MC, Richman DD (2013) New assays for monitoring residual HIV burden in effectively treated individuals. Curr Opin HIV AIDS 8:106–110
4. Strain MC, Lada SM, Luong T et al (2013) Highly precise measurement of HIV DNA by droplet digital PCR. PLoS One. doi:10.1371/journal.pone.0055943
5. Kelly K, Cosman A, Belgrader P et al (2013) Detection of methicillin-resistant staphylococcus

aureus by a duplex droplet digital PCR assay. J Clin Microbiol 51:2033–2039
6. Heredia NJ, Belgrader P, Wang S et al (2012) Droplet Digital™ PCR quantitation of HER2 expression in FFPE breast cancer samples. Methods 59:S20–S23
7. Dodd DW, Gagnon KT, Corey DR (2013) Digital quantitation of potential therapeutic target RNAs. Nucleic Acid Ther 23:188–194
8. Boettger LM, Handsaker RE, Zody MC et al (2012) Structural haplotypes and recent evolution of the human 17q21.31 region. Nat Genet 44:881–885
9. Porensky PN, Mitrpant C, McGovern VL et al (2012) A single administration of morpholino

antisense oligomer rescues spinal muscular atrophy in mouse. Hum Mol Genet 21: 1625–1638

10. Gevensleben H, Garcia-Murillas I, Graeser MK et al (2013) Noninvasive detection of HER2 amplification with plasma DNA digital PCR. Clin Cancer Res 19:3276–3284

11. Nadauld L, Regan JF, Miotke L et al (2012) Quantitative and sensitive detection of cancer genome amplifications from formalin fixed paraffin embedded tumors with droplet digital PCR. Transl Med (Sunnyvale). Doi: pii: 1000107

12. Hindson BJ, Ness KD, Masquelier DA et al (2011) High-throughput droplet digital PCR system for absolute quantitation of DNA copy number. Anal Chem 83:8604–8610

13. Yeh I, von Deimling A, Bastian BC (2013) Clonal BRAF mutations in melanocytic nevi and initiating role of BRAF in melanocytic neoplasia. J Natl Cancer Inst 105: 917–919

14. Morisset D, Stebih D, Milavec M et al (2013) Quantitative analysis of food and feed samples with droplet digital PCR. PLoS One. doi:10.1371/journal.pone.0062583

15. Warren L, Bryder D, Weissman IL et al (2006) Transcription factor profiling in individual hematopoietic progenitors by digital RT-PCR. Proc Natl Acad Sci U S A 103:17807–17812

16. Ottesen EA, Hong JW, Quake SR et al (2006) Microfluidic digital PCR enables multigene analysis of individual environmental bacteria. Science 314:1464–1467

17. Fan HC, Stephen R (2007) Detection of aneuploidy with digital polymerase chain reaction. Anal Chem 79:7576–7579

18. Morrison T, Hurley J, Garcia J et al (2006) Nanoliter high throughput quantitative PCR. Nucleic Acids Res 34:e123

19. Sundberg SO, Wittwer CT, Gao C et al (2010) Spinning disk platform for microfluidic digital polymerase chain reaction. Anal Chem 82:1546–1550

20. Huggett JF, Foy CA, Benes V et al (2013) The digital MIQE guidelines: minimum information for publication of quantitative digital PCR experiments. Clin Chem 59:892–902

mRNA and microRNA Purity and Integrity: The Key to Success in Expression Profiling

Benedikt Kirchner, Vijay Paul, Irmgard Riedmaier, and Michael W. Pfaffl

Abstract

RNA quality control is a crucial step in guaranteeing integer nondegraded RNA and receiving meaningful results in gene expression profiling experiments, using micro-array, RT-qPCR (Reverse-Transcription quantitative PCR), or Next-Generation-Sequencing by RNA-Seq or small-RNA Seq. Therefore, assessment of RNA integrity and purity is very essential prior to gene expression analysis of sample RNA to ensure the accuracy of any downstream applications. RNA samples should be nondegraded or fragmented and free of protein, genomic DNA, nucleases, and enzymatic inhibitors. Herein we describe the current state-of-the-art RNA quality assessment by combining UV/Vis spectrophotometry and microfluidic capillary electrophoresis.

Key words RNA purity, RNA integrity, microRNA (miRNA), mRNA, Microfluidic capillary electrophoresis, UV/Vis spectrophotometer, Denaturing gel electrophoresis, qPCR, MIQE

1 Introduction

Quantification of RNA expression levels serves as a prime indicator of the physiological status of a cell or tissue and plays a central role in a wide variety of life science studies. The purity and integrity of RNA samples were shown to have a direct influence on the outcome of gene expression experiments and may strongly compromise the accuracy of any RNA profile, irrelevant of the method by which it was obtained [1–3]. RNAs are very sensitive molecules, especially compared to DNAs, and are easily fragmented by heat, UV, or the ubiquitous occurring nucleases. In addition contaminants introduced through sloppy lab handling and ineffective sampling or extraction procedures, like salts, phenol, or heparin, were proven to impair downstream reactions and overall affect quantitative gene expression results [4]. Therefore, special care should be taken during each step of RNA preparation (e.g., tissue sampling and storage, RNA extraction, stabilization, and storage)

Roberto Biassoni and Alessandro Raso (eds.), *Quantitative Real-Time PCR: Methods and Protocols*, Methods in Molecular Biology, vol. 1160, DOI 10.1007/978-1-4939-0733-5_5, © Springer Science+Business Media New York 2014

to avoid any contamination or degradation [5–7]. Especially in clinical application, where diagnostic, therapeutic, and prognostic conclusions are drawn and sampling tissues tend to be unique and limited, a reliable and standardized RNA quality control is essential. The necessity of RNA quality control is also highlighted by the MIQE (Minimal Information for Publication of Quantitative Real-Time PCR Experiments) guidelines where it is listed as an essential required element of sample preparation prior to qPCR analysis [8].

RNA quality is defined by a set of criteria that all samples must fulfill in order to obtain comparable and reproducible results. At first, RNA preparations should be free of protein and any enzymatic inhibitors or complexing substances of RT and PCR. Secondly, samples should be nondegraded and free of nucleases. Lastly, contamination with genomic DNA should be excluded [8, 9]. While DNA contamination can be determined easily by negative reverse transcription controls during qPCR analysis, all other parameters should be evaluated prior to that. RNA purity can be assessed most conveniently on a spectrophotometer by measuring the optical density (OD) and comparing the absorption at different wavelengths. Nucleic acids (RNA as well as DNA) have their absorption maxima at 260 nm whereas proteins have their maxima at 280 nm. Additionally, contaminant and background absorption can be measured at 230 and/or 320 nm. An OD260/280 ratio higher than 1.8 is generally viewed as suitable for gene expression profiling [10]. On the other hand the OD260/230 and OD260/320 ratios should be maximized, as no fixed values exist for them since they depend mostly on the used sample tissues and extraction protocols. Preferably, spectrophotometer instruments should be used that do not rely on a cuvette format to avoid positioning errors and excess sample consumption (e.g., NanoDrop, Thermo Fischer Scientific; NanoVue, GE Healthcare; NanoPhotometer, Implen).

To check for enzyme inhibitors a dilution series of the sample in question is quantified via RT-qPCR and correlated against their respective Cq values in a semilogarithmic plot. Noninhibited reactions should exhibit a high linearity (determined by the coefficient of determination, R^2) and qPCR efficiency (determined by slope of the linear regression). Optionally, if using only very small amounts of RNA (samples from, e.g., single cells, laser capture microdissection, or biopsies) a universal inhibition assay such as SPUD can be performed. By measuring a positive qPCR control that lacks homology to any known sequence, in and without the presence of nucleic acid samples any inhibition will be clearly shown in a rise of the corresponding Cq [11].

Various methods have been proposed for the measurement of RNA integrity, but over the last decade microfluidic capillary electrophoresis has emerged as the preferred technology. By combining easy handling even for large numbers of samples and offering the most objective way of assessing the RNA degradation level, instruments such as Agilent Technologies' 2100 Bioanalyzer or Bio-Rad Laboratories' Experion has become the standard for RNA quality control [12]. RNA samples are separated electrophoretically on a microfabricated chip, and fragments are detected via laser-induced fluorescence measurement. Estimation of RNA band sizes and total concentration is achieved by using an RNA ladder as a mass and size standard. Comparable to old fashioned denaturing agarose gel electrophoresis, RNA integrity is mostly determined by the ratio of 28S to 18S rRNA (ribosomal RNA of eukaryotic samples) bands. Ideally, the ratio should be around 2.0, since 28S rRNA has approximately twice the quantity of 18S rRNA, but this is rarely accomplished in practice. Elevated levels of degradation appear as an increased threshold baseline as well as a decreased 28S/18S ratio in the electropherogram [13] (Fig. 1). In addition the instrument's software calculates an objective numerical value based on the rRNA ratio and the occurring RNA bands, ranging from one (almost completely degraded) to ten (intact and nonfragmented). For reliable PCR results a RIN (RNA Integrity Number; Agilent Technologies) or RQI (RNA Quality Index; Bio-Rad Laboratories) of higher than five has been proposed [12]. Although there is conflicting literature data about the correlation of mRNA integrity with 18S or 28S rRNA [14, 15], it is generally believed that mRNA degradation closely resembles that of 28S rRNA. Evaluation of miRNA integrity remains difficult and little is known about the accessibility of miRNAs to degradation processes. However as for mRNA, a significant correlation between miRNA expression data and RIN values was demonstrated [1]. Supplementary information can be gained by the small RNA assay from Agilent Technologies, enabling the separation and analysis of RNA fragments with less than 200 nt and therefore 91 quantifying the absolute concentration of these small RNA fractions and the respective percentage of miRNAs (Fig. 2). Since ongoing RNA degradation, especially longer mRNA species, causes the formation of smaller degraded RNA fragments, it will also be shown as an overrepresentation of the miRNA amount [1].

Thus by combining UV/Vis spectrophotometry and microfluidic capillary electrophoresis we are able to reliably and reproducibly assess the quality of mRNA and miRNA samples with minimal effort and sample consumption. By choosing only biological samples of adequate RNA purity and integrity for gene expression profiling we can now guarantee the correctness and validity of our quantitative results.

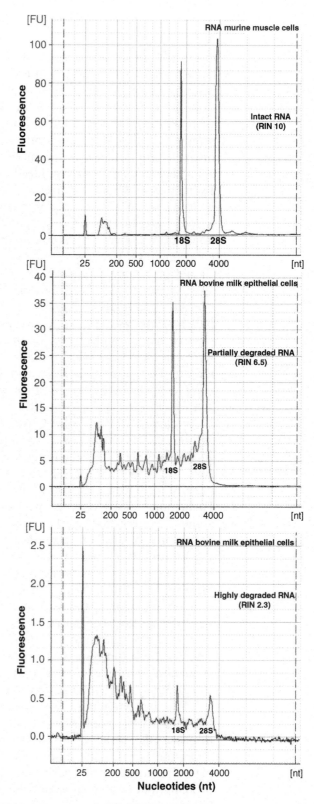

Fig. 1 Electropherograms of total RNA with varying degradation levels (perfect and integer total RNA = RIN 10; intermediate RNA quality with partial degradation = RIN 6.5; and highly degraded RNA = RIN 2.3) using the 2100 Bioanalyzer and the RNA 6000 Nano Kit (Agilent Technologies)

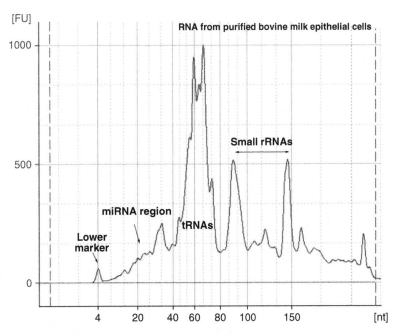

Fig. 2 Electropherogram of a small RNA integrity analysis using the 2100 Bioanalyzer and the Small RNA Kit (Agilent Technologies)

2   Materials

Use only PCR-grade water (DEPC-treated, double distilled, deionized, autoclaved, and free of nucleases) for all preparations and solutions. Prepare and store all reagents at room temperature (unless indicated otherwise). RNA samples for quality control should be kept on ice during all procedures. Diligently follow all waste disposal regulations.

2.1 UV/Vis Spectrophotometry Components

1. NanoDrop 2000 UV/Vis spectrophotometer (Thermo Fisher Scientific, Waltham, USA).
2. NanoDrop 2000 operating software, version 1.4.2.
3. Ethanol for washing, 70 %.
4. Lint-free lab wipes.

2.2 2100 Bioanalyzer Components

All components except **item 1** are manufactured by Agilent Technologies, Santa Clara, USA.

1. Heating block or water bath.
2. 2100 Bioanalyzer.
3. 2100 Expert software for instrument control and data analysis.
4. Chip priming station.
5. Chip vortexer.

6. Agilent RNA 6000 Nano Kit containing

- RNA Nano chips.
- Electrode cleaners.
- Syringe kit.
- RNA 6000 Nano ladder (store in aliquots at –70 °C).
- RNA 6000 Nano dye concentrate (store at 4 °C).
- RNA 6000 Nano gel matrix (store at 4 °C).
- RNA 6000 Nano marker (store at 4 °C).
- Spin filters.
- PCR clean safe lock tubes.

7. Agilent Small RNA Kit containing

- Small RNA chips.
- Electrode cleaners.
- Syringe kit.
- Small RNA ladder (store in aliquots at –70 °C).
- Small RNA dye concentrate (store at 4 °C).
- Small RNA gel matrix (store at 4 °C).
- Small RNA marker (store at 4 °C).
- Small RNA conditioning solution (store at 4 °C).
- Spin filters.
- PCR clean safe lock tubes.

2.3 Denaturing Agarose Gel Electrophoresis

1. 10× MOPS buffer: 0.2 M 3-(N-morpholino) propane sulfonic acid (MOPS), 50 mM sodium acetate, 10 mM ethylenediaminetetraacetic acid (EDTA). Add nuclease-free water to respective final volume and adjust pH to 7.0 with acetic acid or NaOH (prepared in nuclease-free water).

2. 1 % denaturing agarose gel: For 50 ml solution; cook 0.5 g agarose with 5 ml 10× MOPS buffer and 45 ml water until agarose is completely dissolved (*see* **Note 1**). Wait till agarose solution is cooled down to around 40 °C and add 2 ml formaldehyde (*see* **Note 2**) and let it solidify in a gel chamber.

3. Sample buffer (*see* **Note 3**): For 300 μl; add 150 μl formamide, 50 μl 37 % formaldehyde, 30 μl MOPS buffer, 55 μl bromophenol blue in water mixed with 50 % glycerol, 15 μl ethidium bromide stock solution (*see* **Note 4**).

4. RNA marker ranging from 200 to 6,000 nt.

5. Sodium hydroxide 0.1 M.

6. Electrophoresis chamber.

7. UV transilluminator.

2.4 SPUD Assay	1. SPUD amplicon: 5′-AACTTGGCTTTAATGGACCTCCAAT TTTGAGTGTGCACAAGCTATGGAACACCACGTAA GACATAAAACGGCCACATATGGTGCCATGTA AGGATGAATGT-3′.

1. SPUD amplicon: 5′-AACTTGGCTTTAATGGACCTCCAAT TTTGAGTGTGCACAAGCTATGGAACACCACGTAA GACATAAAACGGCCACATATGGTGCCATGTA AGGATGAATGT-3′.

2. SPUD Forward Primer: 5′-AACTTGGCTTTAATGGACCT CCA-3′.

3. SPUD Reverse Primer: 5′-ACATTCATCCTTACATGGCA CCA-3′ 164.

4. SPUD Taqman probe: 5′-FAM-TGCACAAGCTATGGAACA CCACGT-TAMRA-3′.

5. Reverse transcription and qPCR kit of user's choice.

3 Methods

Carry out all procedures at room temperature unless otherwise specified.

3.1 RNA Purity Control: Optical Density Measurement on Nanodrop

1. Clean the sensor plate carefully with 70 % ethanol and dry it with a lint-free wipe.

2. Start the software and choose nucleic acids measurements.

3. Let the instrument initialize itself by pipetting 1.5 μl of water on the sensor plate and measuring it.

4. Wipe the sensor plate clean and apply 1.5 μl of the solution in which your RNA is dissolved to blank the background photospectrum and remove it from the analysis.

5. Once again wipe the sensor plate clean and proceed to measure your RNA samples.

6. Analyze the obtained spectra concerning its absorption at 260, 280, and 230/320 nm with their respective ratios (*see* **Note 5**).

3.2 Total RNA Integrity Control: Microfluidic Capillary Electrophoresis on Agilent 2100 Bioanalyser

1. Denature all RNA samples as well as the RNA 6000 Nano ladder at 68 °C for 2 min and immediately cool them on ice (*see* **Note 6**). For best readouts dilute all RNA samples to a final concentration of 100 ng/μl (*see* **Note 7**).

2. Let the RNA 6000 Nano kit reagents including dye concentrate to equilibrate at room temperature for 30 min in dark and protected from light.

3. Clean the electrodes by pipetting approximately 350 μl of water in the electrode-cleaning chip and putting it in the instrument for 60 s. Repeat it once.

4. Prepare the gel by pipetting 550 μl of RNA 6000 Nano gel matrix into a spin filter and centrifuging it at $1,500 \times g$ for 10 min (*see* **Note 8**).

5. Add 1 µl of dye concentrate to 65 µl of filtered gel in a fresh microcentrifuge tube, vortex it briefly, and centrifuge it at 13,000×g for 10 min (*see* **Note 9**).

6. Put a RNA 6000 Nano chip on the chip priming station and pipette 9 µl of gel-dye mix into the well marked with a white G in a black circle (*see* **Note 10**).

7. Lift the plunger to the 1 ml position, adjust the lever of the clip to top position, and close the priming station.

8. Press down plunger until it is arrested by clip and wait for exactly 30 s.

9. Release the clip and after 5 s slowly pull the plunger to its starting position (*see* **Note 11**).

10. Open the priming station and pipette 9 µl of gel-dye mix in each of the wells marked with a black bold G.

11. Load 5 µl of RNA 6000 Nano marker in all 12 sample wells and the ladder well.

12. Pipette 1 µl of respective RNA sample in each of the 12 sample wells and 1 µl of ladder into the ladder well. Pipette 1 µl of marker in every unused sample well.

13. Vortex the chip for 1 min at 2,400 rpm in the chip vortexer and start the run within 5 min on Agilent 2100 bioanalyser.

14. Analyze the run by checking rRNA ratios, general electropherogram progression, and RIN values.

3.3 Small RNA Integrity Control: Microfluidic Capillary Electrophoresis on Agilent 2100 Bioanalyser

1. Denature all RNA samples as well as the Small RNA ladder at 68 °C for 2 min and immediately cool them down on ice (*see* **Note 6**). For best readouts dilute all RNA samples to a final concentration of 50 ng/µl total RNA.

2. Let all other reagents equilibrate at room temperature for 30 min while protected from light.

3. Clean the electrodes by pipetting approximately 350 µl of water in the electrode cleaning chip and putting it in the instrument for 60 s. Repeat it once.

4. Prepare the gel by pipetting 650 µl (complete volume of one tube) of small RNA gel matrix into a spin filter and centrifuging it at 10,000×g for 15 min (*see* **Note 8**).

5. Add 40 µl of filtered gel to 2 µl of vortexed Small RNA dye concentrate in a new tube, vortex it briefly and centrifuge it at 13,000×g for 10 min (*see* **Note 12**).

6. Put a Small RNA chip on the chip priming station and pipette 9 µl of gel-dye mix into the well marked with a white G in a black circle (*see* **Note 10**).

7. Lift the plunger to the 1 ml position, adjust the lever of the clip to the lowest position and close the priming station.

8. Press down plunger until it is arrested by clip and wait for exactly 60 s.

9. Release the clip and after 5 s slowly pull the plunger to its starting position (*see* **Note 11**).

10. Open the priming station and pipette 9 µl gel-dye mix in each of the wells marked with a black bold G.

11. Slowly pipette 9 µl of small RNA conditioning solution in the well marked with CS.

12. Load 5 µl of small RNA marker in all 11 sample wells and the ladder well.

13. Pipette 1 µl of RNA sample in each of the 11 sample wells and 1 µl of ladder into the ladder well. Pipette 1 µl of marker in every unused sample well.

14. Vortex the chip for 1 min at 2,400 rpm in the chip vortexer and start the run within 5 min on 2100 bioanalyser.

15. Analyze the run by checking relative miRNA content and general electropherogram progression.

3.4 Optional Test for qPCR Inhibitors: SPUD Assay

1. Prepare cDNA solution from 500 ng of total RNA according to the manufacturer instructions.

2. Prepare qPCR master mix according to the manufacturers instruction with 240 nM of each forward and reverse SPUD primer and 200 nM SPUD TaqMan probe.

3. Perform qPCR SPUD assay by measuring the SPUD amplicon in the presence of water (negative control with no inhibitors) and in the presence of your RNA samples. SPUD amplicon should be in the range of 20,000 copies per reaction to ensure high reproducibility and reliability of the fluorescence signal.

4. Analyze amplification plots and obtained Cq's. A higher Cq and lower amplification curve compared to the control sample indicates qPCR inhibitors in your RNA extractions.

3.5 Alternative Method for Total RNA Integrity Measurement: 18S/28S Ratio Agarose Gel Electrophoresis

1. Before preparing the gel, wash the gel chamber for 2 min in 0.1 M NaOH to destroy RNases. Thoroughly flush the chamber with water and let it dry.

2. Prepare the 1 % denaturing agarose gel.

3. Dilute 5 µg of each RNA sample and an appropriate volume of RNA marker with the same volume of sample buffer and incubate it at 65 °C for 10 min to completely denature all RNA.

4. Position the prepared gel in the electrophoresis chamber and fill it with 1× MOPS buffer (1:10 dilution of 10× MOPS buffer with nuclease-free water).

5. Load 10–20 µl (depending of the depth of the gel) of denatured samples in the pockets of the gel and let it run at 60 V

for approximately 1–2 h. Check the progress of the separation via the unspecific bromophenol blue band.

6. After separation RNA is visualized with the UV transilluminator at 256 nm. Determine RNA integrity by comparing rRNA band sizes and fluorescence levels as well as any possible RNA fragments that will appear as an unspecific smear with lower nt sizes.

4 Notes

1. Agarose dissolves at around 36 °C. If you use a microwave for heating, pay special attention to possible boiling retardation.

2. Adding the formaldehyde too early can cause it to evaporate and create toxic fumes.

3. Prepare shortly before loading the samples on the gel to guarantee optimal results.

4. If you are concerned with ethidium bromide toxicity, you could also use a replacement like GelRed (Biotium, Hayward, USA).

5. Pay special attention to any shifts of the maximum absorption caused by impurities. Contaminated samples sometimes appear to have a normal spectrum but at a closer look have their maximum absorption at for example 270 nm.

6. If not denatured, secondary structures of RNA will compromise the size separation during electrophoresis.

7. If you are interested in RNA concentration measurement via 2100 Bioanalyzer do not dilute your samples. Be aware that higher concentrations than 500 ng/µl can impair the performance of the chip and are not recommended by the manufacturer.

8. Filtered gel can be stored at 4 °C for approximately 4 weeks.

9. Gel-dye mix lasts for two chips but should be used on the day of preparation.

10. When pipetting on the chip make sure to pipette directly to the bottom of each well and avoid air bubble formation at all costs.

11. If the plunger is not rising on its own after releasing the clip, check if the priming station was closed properly and if the sealing ring inside the station is undamaged. Incompletely primed chips can sometimes be salvaged by priming them a second time.

12. Careful pipetting is strongly recommended due to high viscosity of the gel.

References

1. Becker C, Hammerle-Fickinger A, Riedmaier I et al (2010) mRNA and microRNA quality control for RT-qPCR analysis. Methods 50:237–243. doi:10.1016/j.ymeth.2010.01.010

2. Fleige S, Pfaffl MW (2006) RNA integrity and the effect on the real-time qRT-PCR performance. Mol Aspects Med 27:126–139

3. Vermeulen J, Prete K, de Lefever S (2011) Measurable impact of RNA quality on gene expression results from quantitative PCR. Nucleic Acids Res 39(9):e63. doi:10.1093/nar/gkr065

4. Nolan T, Hands RE, Bustin SA (2006) Quantification of mRNA using real-time RT-PCR. Nat Protoc 1:1559–1582

5. Pérez-Novo CA, Claeys C, Speleman F et al (2005) Impact of RNA quality on reference gene expression stability. BioTechniques 39:54–56

6. Holland NT, Smith MT, Eskenazi B et al (2003) Biological sample collection and processing for molecular epidemiological studies. Mutat Res 543:217–234

7. Schoor O, Weinschenk T, Hennenlotter J et al (2003) Moderate degradation does not preclude microarray analysis of small amounts of RNA. BioTechniques 35(1192–1196):1198–1201

8. Bustin SA, Benes V, Garson JA et al (2009) The MIQE guidelines: minimum information for publication of quantitative real-time PCR experiments. Clin Chem 55:611–622

9. Pfaffl MW (2005) Nucleic acids: mRNA identification and quantification. In: Worsfold P, Townshend A, Poole CF (eds) Encyclopedia of analytical science, 2nd edn. Elsevier Academic Press, Amsterdam, Boston, pp 417–426

10. Manchester KL (1996) Use of UV methods for measurement of protein and nucleic acid concentrations. BioTechniques 20:968–970

11. Nolan T, Hands RE, Ogunkolade W et al (2006) SPUD: a quantitative PCR assay for the detection of inhibitors in nucleic acid preparations. Anal Biochem 351:308–310

12. Fleige S, Walf V, Huch S et al (2006) Comparison of relative mRNA quantification models and the impact of RNA integrity in quantitative real-time RT-PCR. Biotechnol Lett 28:1601–1613

13. Mueller O, Lightfoot S, Schroeder A et al (2006) The RIN: an RNA integrity number for assigning integrity values to RNA measurements. BMC Mol Biol 7:3. doi:10.1186/1471-2199-7-3

14. Miller CL, Diglisic S, Leister F et al (2004) Evaluating RNA status for RT-PCR in extracts of postmortem human brain tissue. BioTechniques 36:628–633

15. Santiago TC, Purvis IJ, Bettany AJ et al (1986) The relationship between mRNA stability and length in Saccharomyces cerevisiae. Nucleic Acids Res 14:8347–8360

Chapter 6

Mediator Probe PCR: Detection of Real-Time PCR by Label-Free Probes and a Universal Fluorogenic Reporter

Simon Wadle, Stefanie Rubenwolf, Michael Lehnert, Bernd Faltin, Manfred Weidmann, Frank Hufert, Roland Zengerle, and Felix von Stetten

Abstract

Mediator probe PCR (MP PCR) is a novel detection format for real-time nucleic acid analysis. Label-free mediator probes (MP) and fluorogenic universal reporter (UR) oligonucleotides are combined to accomplish signal generation. Compared to conventional hydrolysis probe PCRs costs can thus be saved by using the same fluorogenic UR for signal generation in different assays. This tutorial provides a practical guideline to MP and UR design. MP design rules are very similar to those of hydrolysis probes. The major difference is in the replacement of the fluorophore and quencher by one UR-specific sequence tag, the mediator. Further protocols for the setup of reactions, to detect either DNA or RNA targets with clinical diagnostic target detection as models, are explained. Ready to use designs for URs are suggested and guidelines for their de novo design are provided as well, including a protocol for UR signal generation characterization.

Key words Mediator probe PCR, Mediator probe, Universal reporter, Universal sequence-dependent nucleic acid detection, Real-time PCR

1 Introduction

In real-time nucleic acid analysis, several probe-based detection formats are well established, most prominently hydrolysis probe, molecular beacon, or hybridization probe PCR [1–3]. These dual-labeled probe formats however bare disadvantages in particular applications in research or assay development: Typically 2–4 design iterations per probe are used to identify those with optimal performance [4–7]. Selection of optimum fluorescent labels and quencher molecules required for fluorescence resonance energy transfer (FRET)-based detection is an empirical process as well. Furthermore these molecules often show a large lot-to-lot variance in their

Roberto Biassoni and Alessandro Raso (eds.), *Quantitative Real-Time PCR: Methods and Protocols*, Methods in Molecular Biology, vol. 1160, DOI 10.1007/978-1-4939-0733-5_6, © Springer Science+Business Media New York 2014

fluorescence emission or quenching efficiency impairing the comparability of data [8–10]. Restrictions in the possible length of a dual-labeled probe and requirements for sequence composition (5'-end bases, GC content between 50 and 60 %) reduce the flexibility in probe design and design of applications [8, 11].

To overcome this problem, a number of nucleic acid detection formats have been developed that use universal fluorogenic reporter oligonucleotides. These can be applied for the detection of different nucleic acid sequences and thus avoid the mentioned cost-, signal generation-, and sequence composition issues. Available methods for the universal sequence-dependent detection (USD) of nucleic acids have been recently reviewed by Faltin et al. [12] and more are still upcoming [13–15]. This tutorial is dedicated to mediator probe PCR (MP PCR) [16] representing one of these USD approaches. MP PCR excels especially by high detection selectivity towards specific over unspecific amplification products.

Sequence-specific detection is accomplished by the label-free primary probe, the mediator probe (MP). The MP consists of a 3'-terminal sequence region, which hybridizes the target nucleic acid sequence during amplification and of a 5'-terminal sequence region, the generic mediator region. For real-time signal generation, MP PCR combines two reactions in one nucleic acid amplification cycle (Fig. 1): (1) The MP is cleaved during primer extension resulting in the release of the mediator. Cleavage site is 3'-off the 5'-terminal base of the target-specific region (Fig. 2). (2) Signal generation is triggered by the hybridization of the mediator to a universal fluorogenic reporter (UR) oligonucleotide (Fig. 3).

Quenching and fluorescence emission efficiencies for the UR have to be optimized just once and then one or more UR can be repeatedly used in different assays by combination with a large variety of target-specific MPs. The latter are flexible in sequence length, GC content, and sequence composition in general, since no fluorescence efficiency considerations must be regarded.

MP PCR assay setup and almost the entire oligonucleotide (primer and target-specific probe region) design are analogous to hydrolysis probe PCRs (HP PCR/TaqMan PCR). Materials and methods considerations of HP PCRs have been described very detailed before [9, 17, 18] and can be adopted as basis for MP PCR assay design.

In the subsequent Subheading 2 we introduce the biochemicals required to perform a MP PCR or a reverse transcription MP PCR (RT-MP PCR). This is followed by Subheading 3 that describes the design of MPs and URs. Further, the experimental workflow is presented from reagent setup to data acquisition with common real-time PCR thermocyclers for: (1) DNA detection in MP PCR or (2) duplex MP PCRs, (3) RNA detection using RT-MP PCR, and (4) the performance characterization of custom-designed URs.

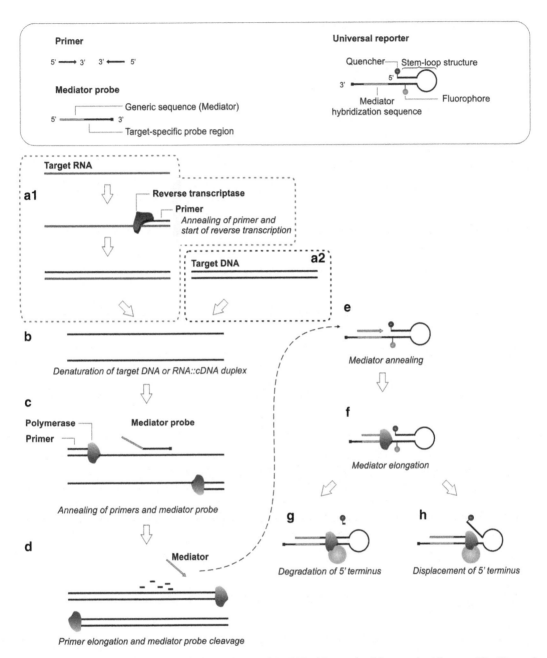

Fig. 1 Schematic illustration of the mediator probe (RT-) PCR. Oligonucleotides required for amplification and detection are shown in the *upper box*. Amplification and detection are shown in steps *A* to *H*. The nucleic acid target can be either RNA (*a1*) or DNA (*a2*). For RNA, reverse transcription into cDNA is required prior to amplification. Next, the reversely transcribed RNA or the DNA target is denatured at elevated temperatures (*b*). Annealing of mediator probe and primer molecules (*c*). The 5′ portion of the mediator probe does not anneal to the target. Primer extension and cleavage of mediator probe (*d*). With each target duplication one mediator is released to the bulk solution. Subsequently, the mediator anneals to the universal reporter (*e*). Mediator elongation (*f*) leads to de-quenching of the fluorophore induced either by sequential degradation of the 5′ terminus and release of the quencher moiety (*g*) or displacement of the 5′ terminus and unfolding of the stem–loop structure (*h*). Both ways contribute to signal generation. All reaction steps take place within one thermocycle. (Modified after Faltin [12] with permission from AACC)

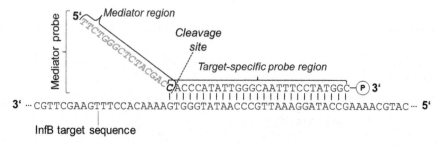

Fig. 2 Alignment of MP with target binding site. The target-specific region of the InfB MP is reverse complementary to the corresponding hybridization site of the target. The mediator region (indicated in *grey*) comprises
a sequence, which does not bind the target at the corresponding reaction conditions. Upon annealing of the
MP to the target the 5′-terminal cytosine of the target-specific region (indicated in *bold & italics*) will be
cleaved off by the polymerase's 5′–3′ nuclease activity together with the mediator region, resulting in release
of a 17 nt mediator. (P) = phosphate blocking moiety

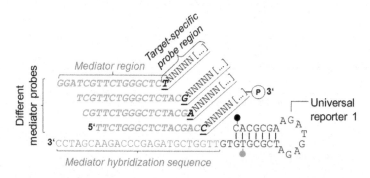

Fig. 3 Alignment of alternative MP sequences (uncleaved = initial status) with UR1. All MPs are depicted in
5′–3′ orientation (*left* to *right*). The mediator region comprising the *grey* sequence noncomplementary to the
target and the first binding nucleobase at the target-specific region (thymidine, guanosine, adenosine, or cytosine; all *bold* and *underlined*) must hybridize the UR under the corresponding reaction conditions. Except of this
base the remaining target-specific probe region (NNNNN [...], *underlined*) does not interact with the UR. The
3′-phosphate group (P) is exemplarily given for only one MP. *Green* and *black spots* at UR correspond to the
fluorophore and quencher moiety, respectively

2   Materials

2.1   Nucleic Acid Template

DNA and RNA can be used as template for MP PCR and RT-MP
PCR, respectively, as in conventional real-time PCR assays.
Examples used in the tutorial including storage conditions are the
following.

1. DNA: Lambda phage DNA (New England Biolabs, Germany,
   Cat. No. N3011S).

2. DNA: Human Adenovirus species B serotype 7 (HAdV B7,
   140 bp PCR product of hexon gene in cloning vector pCRII):
   also Lambda phage DNA was diluted in this buffer.

3. RNA: Influenza B virus (InfB, in vitro RNA transcript of synthetic DNA standard of the hemagglutinin gene (HA gene) in cloning vector pCRII). RNA was diluted in RNA storage solution (Ambion, USA, Cat. No. AM7000) and used as amplification standard.

2.2 Primers, Mediator Probes, and Universal Reporter

This tutorial provides design rules for (RT-) MP PCR primers, MPs, and URs. If available, also established PCR primers and target-specific probe regions (e.g., originally designed for hydrolysis probe PCRs) can be used.

Primers MP and UR designed for the experiments described in this tutorial are listed (Table 1). They specifically amplify and detect the template sequences listed in Subheading 2.1. HAdV B7 and InfB primer and probe sequences were designed to match highly conserved regions. Oligonucleotide synthesis was performed by biomers.net GmbH, Ulm, Germany.

For storage of the oligonucleotides, we recommend preparation of 100 μM stock solution and 10 μM working solution aliquots in 1× TE buffer (pH 7.5–8.0) or H_2O and storage at –20 °C. Repeated freeze–thaw cycles should be avoided.

2.3 Enzymes and Buffers

MP PCR requires a polymerase with 5′–3′ nuclease activity. The use of HotStart polymerases can avoid unwanted initiation of non-specific amplification reactions during PCR setup.

For the experiments described in this tutorial the following reagents have been used.

1. DNA detection by MP PCR (*see* Subheadings 3.1.1 and 3.1.3) and UR characterization (*see* Subheading 3.2.1): HotStar TaqPlus polymerase (5 U/μl, Qiagen, Hilden, Germany, Cat. No. 203603) with reaction buffer compounds (GenoID 2× PCR buffer (GenoID, Hungary), 10 mM dNTPs (Qiagen, provided with polymerase), 100× BSA (New England Biolabs, USA)).

2. RNA-detection by MP RT-PCR (*see* Subheading 3.1.1): Qiagen QuantiTect Multiplex RT-PCR NR Kit (Qiagen, Cat. No. 204843).

   Apart from these the following reagents have been successfully tested to perform (RT-) MP PCR:

   - Qiagen HotStar TaqPlus polymerase with reaction buffer compounds (Qiagen 10× PCR buffer, 10 mM dNTPs) (Qiagen, Cat. No. 203603).

   - Qiagen HotStar TaqPlus Mastermix Kit (Qiagen, Cat. No. 203443).

   - Thermo Scientific DyNAmo Flash Probe qPCR Kit (Thermo Scientific, USA, Cat. No. F-455).

   - Roche LightCycler 480 RNA Master Hydrolysis Probes (Roche, Germany, Cat. No. 04991885001).

Table 1
Oligonucleotide sequences used for model (RT-) MP PCRs described in this tutorial

Oligonucleotide or target name	Description	Sequence (5'–3')	Modification		Length (nt)
			5'	3'	
Universal Reporter 01	UR1—Cy5[a]	CACGCG*A*A*GATGAGATCGCG(dT-Cy5) GT*GTTGGTCGTGAGCCCAGAACGATCC*	BHQ-2	C₃ Spacer	49
Universal Reporter 02	UR2—FAM[a]	CACCGG*C*C*AAGACGCGCCGG(dT-6FAM) GT*GTTCACTGACCGAACTGGAGCA*	BHQ-1	C₃ Spacer	45
Mediator for UR1	Mediator UR1	TTCTGGGCTCTACGACC	–	–	17
Lambda phage DNA	Forward primer (Fw Lambda)	GGGATCATTGGGTACTGTGG	–	–	20
	Reverse primer (Rv Lambda)	CAGACTTGGGGGGTGATGAGT	–	–	20
	Mediator probe[b] (MP Lambda) – UR2-FAM	***GCTCCAGTTCGGTCAGTG|TAAA*** AACACCTGACGCGCTATCCCTG	–	PH	43
	Forward primer (Fw InfB)	GGATTAAATAAAAGCAAGCCTTAC	–	–	24
Influenza B virus (InfB) Hemaglutinin gene	Reverse primer (Rv InfB)	CAGCAATAGCTCCGAAGAAAC	–	–	21
	Hydrolysis probe (HP InfB)	CACCCATATTGGGCAATTTCCTATGGC	6-FAM	BHQ-1	27
	Mediator probe[b] (MP InfB) – UR1-Cy5	*TTCTGGGCTCTACGACC|ACCC* ATATTGGGCAATTTCCTATGGC	–	PH	43
Human Adenovirus 7 species B Hexon gene	Forward primer 1(Fw1 HAdV)	CATGACTTTTGAGGTGGATCCCA	–	–	23
	Forward primer 2 (Fw2 HAdV)	GAATTTCGAAGTCGACCC	–	–	18
	Reverse primer (Rv HAdV)	CCGGCCGAGAAGGGTGTGCGCAGGTA	–	–	26
	Mediator probe[b] (MP HAdV) – UR1-Cy5	*TTCTGGGCTCTACGACCA|CCAGC* CACACGCGGGCG	–	PH	35

[a]"*" = phosphorothioate, underlined indicates stem, italics + bold indicates mediator hybridization site

[b]italics + bold indicates Mediator region, underlined indicates target-specific probe region; PH = phosphate moiety, dashed vertical line indicates MP cleavage site

2.4 Real-Time PCR Thermocycler

To perform MP PCR a conventional real-time PCR thermocycler can be used. For the experiment described in this tutorial a Rotor-Gene Q (Qiagen) was used.

- Further real-time thermocyclers have been tested successfully to perform MP PCR: Roche Light-Cycler 480.
- Applied Biosystems 7900HT.
- Siemens Versant.

3 Methods

Sample protocols for the design and setup of MP PCRs are provided in this chapter, covering a range of different applications.

3.1 MP PCR and RT-MP PCR Design

Considerations for the design of primer and mediator probes (MP) are provided for a ready-to-use UR design (UR1, Table 1). As a model for MP PCR the primer and MP designs for HAdV B7 detection, and for RT-MP PCR the corresponding oligonucleotide designs for InfB detection are presented, respectively.

Selection of Primer and Probe Sequences

For the given target sequence to be detected both, primer- and target-specific probe region sequences of the MP (Fig. 1), can be selected using supportive software tools, such as primer3 [19], PrimerQuest [20], VisualOMP™ [21], and others. For MP PCR or RT-MP PCR (including the two models HAdV B7 and InfB detection, respectively) the following settings should be applied independent of the design tools used:

- Product size range: 100–200 bp.
- Melting temperature (T_m) of primers: 58–65 °C (depending on PCR reaction buffer).
- T_m of target-specific probe region: minimum 5 °C above T_m of primers. Uncleaved MPs must bind the target sequence before binding of primer to the target and before binding of the mediator region to the UR. Thus the T_m of the primer (and mediator region, see below) must be minimum 5 °C below the T_m of the target-specific probe region.
- Length and GC content of primers and target-specific probe region: 15–30 nt; GC 30–70 %.
- *Consideration for RT MP PCR:*
  RNA is more prone to secondary structure generation then DNA. Therefore RNAfold [22] should first be used to find regions of the target sequence, which show a high degree of folding. These regions should be defined as "excluded regions" in the software tools for primer and probe selection. As primers for RNA into cDNA conversion we recommend sequence-specific primers rather than oligo(dT) primers or random hexamers, due to most specific synthesis of cDNA [23].

The described design settings resulted in the primer sequences and the target-specific probe region sequences of the MPs (underlined) as given (Table 1).

MP Design

Design of the target-specific probe region must be adapted at the 5'- and 3'-terminal end to result in a complete MP.

At the 5'-terminus the mediator region is added. The mediator hybridization sequence of the UR (in this example of UR1) defines the potential mediator region to be used as:

5'-GGATCGTTCTGGGCTCTACGACCAA-3' (25 nt).

The mediator region selected out of this 25 nt long sequence should have the same T_m as the primer with the lowest T_m. T_m of the mediator region must be minimum 5 °C below the T_m of the target-specific probe region (see above). Depending on the 5'-terminal base of the target-specific probe region, one of four different mediators depicted in Fig. 3 could be selected to enable MP-UR hybridization as shown. All mediator regions shown are 17 nt long, which was found being cleaved well by the Taq polymerases used.

To the 3'-terminal end of the MP a blocking moiety is added which prevents extension by the polymerase (Fig. 1). As standard we use a phosphate group, but other groups, avoiding extension, such as biotin, carbohydrate spacer (e.g., C3-spacer), or C6-aminolinker can be used as well.

For both examples, the DNA (HAdV B7) and the RNA (InfB) detection, the resulting MP HAdV and MP InfB are given in Table 1.

Hybridization and Folding Check

Unwanted interactions such as hybridization of primers with the MP or the UR, respectively, must be investigated in silico at the corresponding reaction conditions. Pairwise alignment software [24] can be used. In case of stable hybridization (especially of the 3'-termini of primers), i.e., the Gibbs free energy (ΔG) of the dimer is <0 kcal mol^{-1} at the given reaction conditions, the interacting primer should be replaced by another one suggested by the primer design software (*see* Subheading 3.2.1).

Furthermore, the MP-target sequence and MP-UR interactions must be checked at the corresponding reaction conditions, e.g., using the same alignment tool as for the primer interaction check. The MP must only interact with the target binding site and the UR as described:

- The target-specific probe region hybridizes to the corresponding target site, and a 16 nt mediator region at the 5' terminus of the probe does not hybridize (Fig. 2).

- The 17 nt mediator region must specifically anneal to the mediator hybridization sequence of the UR and not to any other part of the UR (Fig. 3).

In case of unwanted interactions, i.e., unspecific folding or hybridization of the MP at the target or UR other than indicated in Figs. 2 or 3, an alternative mediator region can be selected according to the guidelines provided (*see* Subheading 3.1.1). If unwanted interactions are still present, the target-specific probe region can be switched by a few bases in 5'- or 3'-terminal orientation alongside the template. The hybridization and folding check must then be repeated for the novel MP design (*see* **Note 1**).

3.2 Experimental Setup of MP PCR and RT-MP PCR

Performing DNA or RNA detection using the oligonucleotides designed (*see* Subheading 3.1). In detail, two model setups are shown for DNA (HAdV B7) or RNA (InfB) detection. In the latter model, the RT step for RNA in cDNA conversion is performed in the same reaction tube with the subsequent PCR. Alternatively RT can be performed in an isolated reaction with subsequent mixing of the cDNA product with the MP PCR reagents.

MP PCR Mastermix

- DNA detection (HAdV B7).
  All reagents are mixed in DNase/RNase-free PCR tubes. During reaction setup all components are kept on ice (~4 °C). The final volume of the mastermix is 200 µl containing the components:
  - Qiagen HotStar TaqPlus Polymerase (0.1 U/µl).
  - GenoID PCR buffer (1×).
  - NEB BSA (6×).
  - dNTPs (0.2 mM each).
  - "Fw1 HAdV" (300 nM).
  - "Fw2 HAdV" (300 nM).
  - "Rv HAdV" primer (300 nM).
  - "MP HAdV" (150 nM).
  - "UR1" (50 nM).
  - DNase/RNase-free water is used to adjust concentrations of the mastermix components.
- RNA detection (InfB).
  Mixing of the 200 µl volume mastermix prepared is performed as for DNA with the following reagents:
  - Qiagen QuantiTect Multiplex RT-PCR NR Kit (1×).
  - "Fw InfB" (300 nM).
  - "Rv InfB" primer (300 nM).
  - "MP InfB" (150 nM).
  - "UR1" (50 nM).
  - DNase/RNase-free water is used to adjust concentrations.

– As a reference the same mastermix is prepared in a hydrolysis probe format, in which the MP InfB is replaced by the HP InfB (150 nM) and no UR is added (Table 1).

Addition of Nucleic Acid Template

To analyze different target nucleic acid dilutions, the mastermix (either for DNA or RNA detection) is distributed into five aliquots of 36.6 μl each (Rotor-Gene PCR tubes, 0.2 ml). Then 4.4 μl of each dilution (10^5–10^2 copies per μl) of either the HAdV B7 DNA or the InfB RNA are added to the mastermix aliquots 1–4. The no template control (NTC) contained 4.4 μl H_2O in place of nucleic acid template (aliquot 5). Then each of the supplemented aliquots is distributed to four tubes (Rotor-Gene PCR tube strips, 0.1 ml), resulting in four technical replicates of 10 μl each. The mentioned volumes consider a pipetting error of 10 % (v/v).

Thermocycling and Readout

After sealing, the prepared Rotor-Gene PCR tube strips are ready for use with the following protocols to run real-time (RT-) MP PCR in a Rotor-Gene Q (Qiagen).

- DNA detection (HAdV B7).
  This MP PCR protocol can be used for detection of one single HAdV B7 DNA target or for duplex detection demonstrated in Subheading 3.3:
  – Initial denaturation and hot start: 5 min at 95 °C.
  – 50× cycling: 15 s at 95 °C and 45 s at 60 °C.
  – Fluorescence acquisition with readout channel red (Cy5) at the end of each 60 °C cycling step (*see* **Note 2**).
- RNA detection (InfB).
  – Reverse transcription: 20 min at 50 °C.
  – Initial denaturation and hot start: 15 min at 95 °C.
  – 50× cycling: 15 s at 95 °C and 45 s at 60 °C.
  – Fluorescence acquisition with readout channel red (Cy5) at the end of each 60 °C cycling step (Fig. 4a, b) (*see* **Note 2**).

3.3 Duplex MP PCR Design

The primer and MP design with two given URs to detect an analyte sequence of interest ("analyte," detected with UR1) and an internal amplification control ("IAC," detected with UR2) is described. As a design model we present the primer and MP sequences of the duplex MP PCR for HAdV B7 detection with co-amplification of lambda phage DNA (experimental setup described in Subheading 3.4).

Selection of Primer and Probe Sequences

Selection of the analyte and IAC sequences to be detected as well as the design of primer and the target-specific probe regions of the MP are analogous to the design described in Subheading 3.1.1 for single targets. In order to obtain similar oligonucleotide annealing

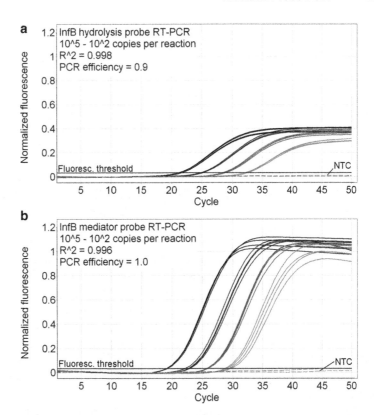

Fig. 4 Real-time fluorescence data of InfB hydrolysis probe (**a**) and mediator probe (**b**) RT-PCR. Both methods show typical sigmoidal fluorescence curves according to the exponential PCR kinetics during amplification of different dilutions of InfB RNA (10^5–10^2 copies per reaction). Linearity (R^2) of the RNA input concentration with the back-calculated RNA concentrations and efficiency were determined from standard curve data (plot of C_q values versus known concentration, not shown). In this example, linearity is comparable for both methods, whereas efficiency of the reaction is slightly higher for MP (RT-) PCR. NTC: no template control

efficiencies for the two different targets, the $T_{m\,(primer)}$ of all primers used and the $T_{m\,(probe)}$ of the target-specific probe regions of the MPs must be in the same range, respectively. To exclude unwanted hybridization of primers and probes pairwise alignment as performed for hybridization check in Subheading 3.1.3 is recommended. In case of primer–dimers or primer–probe–dimers redesign of the oligonucleotides might be necessary.

MP Design for UR1 and UR2

- UR1. The design of the analyte (i.e., HAdV B7)-specific MP exactly follows the steps described in 3.3. The resulting model MP is shown in Table 1.

- UR2. The design of the IAC (i.e., Lambda phage)-specific MP also follows the steps described in Subheading 3.1.2, however, using the following potential mediator region (reverse complementary to mediator hybridization sequence of UR2): 5′-TGCTCCAGTTCGGTCAGTGAAC-3′.

For the T_m of the mediator region selected out of this 25 nt long sequence the same is true as described in Subheading 3.1. According to the 5′-terminal base of the target-specific probe region one of the following four mediator regions could be selected similar to the illustration in Fig. 3:

G as 5′-terminal base: 5′ CTCCAGTTCGGTCAGT**G**
C as 5′-terminal base: 5′ CAGTTCGGTCAGTGAA**C**
A as 5′-terminal base: 5′ TCCAGTTCGGTCAGTG**A**
T as 5′-terminal base: 5′ GCTCCAGTTCGGTCAG**T**

As before, a blocking moiety must be added to the 3′-terminal. The resulting model MP design used for lambda phage DNA (IAC) detection is also shown in Table 1.

Hybridization and Folding Check

All primers, the two MPs and UR1 and 2 must be checked for unwanted hybridization or unintended folding as described in Subheading 3.1.3.

In case of unintended interaction the mediator regions can be redesigned based on the mediator hybridization sequences provided for UR1 and 2 according to the guidelines provided in Subheading 3.1.2 (*see* **Notes 1** and **2**).

3.4 Experimental Setup of Duplex MP PCR

Setup of the duplex MP PCR is analogous to the three steps presented in Subheading 3.2 for single DNA targets. As an example, we present the experimental steps for the duplex MP PCR to detect HAdV B7 (analyte) with co-amplification of lambda phage DNA (IAC).

Duplex MP PCR Mastermix

The same mastermix for HAdV B7 DNA (*see* Subheading 3.2.3) is used augmented by the following oligonucleotides for detection of lambda phage DNA.

- "Fw Lambda" (300 nM).
- "Rv Lambda" (300 nM).
- "MP Lambda" (150 nM).
- "UR2" (50 nM).
- Lambda phage DNA (10^4 copies per 10 μl final reaction).

Addition of Nucleic Acid Template

To analyze different analyte nucleic acid dilutions, the mastermix is distributed to five aliquots of 29.7 μl each (Rotor-Gene PCR tubes, 0.2 ml). Then 3.3 μl of each HAdV B7 DNA dilution (10^5–10^2 copies per μl) is added to the mastermix aliquots 1–4, and 3.3 μl of H_2O is added to mastermix aliquot 5 serving as no template control (NTC). Each of the supplemented aliquots is then divided into 4×8 μl technical replicates in 0.1 ml Rotor-Gene PCR tube strips. The mentioned volumes consider a pipetting error of 10 % (v/v).

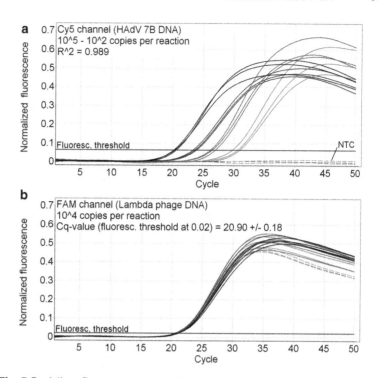

Fig. 5 Real-time fluorescence data of HAdV B7 (**a**) and Lambda phage DNA (**b**) co-detection by duplex MP PCR. Serial dilutions of HAdV B7 DNA (target sequence) were co-amplified with 10⁴ copies of Lambda phage DNA (internal amplification control). Quantitative discrimination of the single target sequence dilutions is possible in the depicted concentration range, whereas the internal amplification control (IAC) shows essentially the same amplification result in all reactions. NTC: no template control

After sealing, the prepared Rotor-Gene PCR tube strips are ready for use in duplex MP PCR with the following protocol in a Rotor-Gene Q (Qiagen).

Real-Time Duplex MP PCR Protocol

The same thermocycling protocol as presented in Subheading 3.2.3 for HAdV B7 detection is used; however, readout is performed for two fluorescence channels: red (for UR1 Cy5 fluorescence readout) and green (for UR2 FAM fluorescence readout) (*see* **Note 2**). Results of the real-time duplex MP PCR are presented in Fig. 5a, b (*see* **Note 3**).

3.5 De Novo UR Design

New URs either as alternatives or augmentation to the presented UR1 or UR2 can be designed. This can also increase the multiplexing degree of MP PCRs by increasing the number of available different mediator hybridization sites. In the following section all relevant guidelines for de novo UR design are provided. An overview on the URs functional regions is provided (Fig. 6) based on UR1 as a tutorial example.

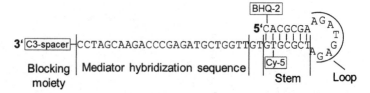

Fig. 6 UR1 with three functional regions. Description of regions in 5′–3′ orientation: (1) A stem–loop structure enables formation of a contact fluorescence quenching pair with a fluorophore moiety (here Cy5) and a quencher moiety at the 5′-terminal end of the oligonucleotide (here BHQ-2). (2) A region for mediator hybridization is the reverse complementary sequence to any applied mediators. (3) Unwanted extension of the 3′-terminal sequence end by polymerase activity is avoided by addition of a blocking moiety (here C3-spacer)

Stem–Loop Region Design

The 5′-terminal stem–loop structure enables efficient contact quenching between the fluorophore and the quencher moiety. The loop has to be a sequence of 6–10 nt in length (e.g., 8 nt for UR1), which does not contain self-complementary regions. Sequence length of the stem should be 6–8 bp. The double-stranded stem must be stable at 70 °C (Gibbs free energy ΔG <0 kcal mol^{-1}; e.g., −3.42 kcal mol^{-1} for UR1). To check the stability of the stem–loop secondary structure (Fig. 6), the free software engine RNAfold [22] can be used with these settings: DNA parameters, no dangling end energies, temperature = 70 °C.

Fluorescence Modifications

At the 5′-terminal end either the fluorophore or quencher moiety can be placed. In the example of UR1 it is the BHQ-2 quencher. To generate an FRET pair the corresponding quencher or fluorophore moiety, respectively, is placed at the nucleobase complementary to the 5′-terminal end inside the stem region or next to the complementary nucleobase as it is the case for UR1. The combination of fluorophore and quencher needs to be in accordance with the capabilities of the synthesis company as well as the readout capabilities of the applied real-time PCR thermocycler (*see* **Note 4**). Due to a duplex destabilizing effect of internal modifications, the quencher or fluorophore moiety must be coupled to the base via a linker and not to the sugar in the backbone. In our example (Fig. 6), Cy5 is linked to the thymine base at internal sequence position 21 via a C6-aminolinker for UR1.

Blocking Moiety

Different blocking moieties can be placed at the 3′-terminal end of the UR: A C3-spacer (used in our example of UR1), a phosphate, a C6-aminolinker, or any other blocking group that can avoid extension of the URs 3′ end by the polymerase.

Mediator Hybridization Sequence (MHS) Design

Design of the MHS starts with generation of a random sequence. This can be supported by software tools, such as the "Random DNA Sequence Generator" [25]. The length and GC content of the MHS should be 20–30 nt, and 50–60 %, respectively. Furthermore, the 5′-terminal end of the MHS should contain all four bases (A G C T) in varying sequence to allow for maximum flexibility when selecting the 3′-terminal base of the mediator (Fig. 3). As the MHS determines the mediator region, which acts as a primer, the MHS should—at the given reaction conditions—neither shows any stable hybridization with any nucleic acid sequences expected in the sample nor with any primers and target-specific probe regions involved in the reaction. Therefore the generated random sequence should be analyzed via Nucleotide-BLAST against relevant genome databases [24] as well as a hybridization check as explained in Subheading 3.1.3 should be applied.

Folding Check

To enable functionality of the UR, its secondary structure as depicted in Fig. 6 needs to be stable at 70 °C. This can be checked using RNAfold with the settings given in Subheading 3.5.1. In case of unintended foldings, repetition of the MHS design (and subsequent folding checks) will lead to a suitable UR structure.

3.6 Experimental UR Characterization

A test setup to characterize functionality of the designed UR is described exemplary based on UR1 (Table 1).

Mastermix Setup

The same mastermix as described for HAdV B7 DNA detection (*see* Subheading 3.2.1) by MP PCR is prepared with the following adaptions: The final volume is 63 µl and neither primers nor MP are added.

Aliquoting and Mediator Addition

The mastermix (which contains UR1) is distributed into two volumes of 29.7 µl which are pipetted into two Rotor-Gene tubes (0.2 ml). One of the aliquots is supplemented with 3.3 µl of "Mediator UR1" (Table 1) (final concentration 125 nM), and the other one with 3.3 µl of H$_2$O. Both supplemented aliquots are distributed into 3 × 10 µl technical replicates (Rotor-Gene PCR tube strips, 0.1 ml).

The prepared Rotor-Gene PCR tube strips are ready for use with the following protocol in a Rotor-Gene Q (Qiagen).

UR Characterization Protocol

Three thermocycling and readout steps are required to characterize signal generation of UR (Fig. 7).

- Initial incubation at constant temperature for background readout:

    10× cycling: 10 s at 60 °C with fluorescence acquisition at the end of each cycle using readout channel red (Cy5).

- Initial denaturation and hot start activation of polymerase: 5 min at 95 °C.

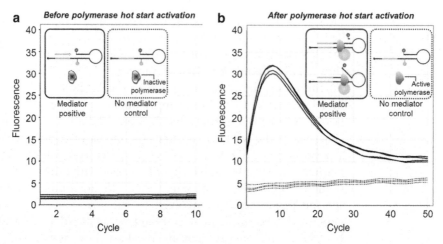

Fig. 7 Characterization of UR1. Fluorescence signal increase can be observed upon addition of "Mediator UR1" and hot-start activation of polymerase (*solid curves*). (**a**) "No mediator control" is used to determine the background signal (*dotted curves*). (**b**) Signal drop after UR activation (cycle 10 onwards) is due to the competition of refolding of UR and extended mediator and the signal generation reaction

- Activation of UR by extension of the mediator and detection of fluorescence signal:
  - 45× cycling of 5 s at 95 °C and 50 s at 60 °C with fluorescence acquisition at the end of each 60 °C step using readout channel red (Cy5).

Data Analysis for UR Characterization

The following quotient gives the effective signal increase upon mediator–extension (Signal$_{Mediator}$) at the UR:

Equation 1

$$\text{Signal}_{\text{Mediator}} = \frac{\text{MF} + \text{cyc}(\text{B}) \, / \, \text{MF} + \text{cyc}(\text{A})}{\text{MF} - \text{cyc}(\text{B}) \, / \, \text{MF} - \text{cyc}(\text{A})}$$

Equation 2

$$\text{ErrorSignal}_{\text{Mediator}} = \sqrt{\left(\frac{\text{SD}_{\text{MF}+\text{cyc}(\text{B})}}{\text{MF}+_{\text{cyc}(\text{B})}}\right)^2 + \left(\frac{\text{SD}_{\text{MF}+\text{cyc}(\text{A})}}{\text{MF}+_{\text{cyc}(\text{A})}}\right)^2 + \left(\frac{\text{SD}_{\text{MF}-\text{cyc}(\text{B})}}{\text{MF}-_{\text{cyc}(\text{B})}}\right)^2 + \left(\frac{\text{SD}_{\text{MF}-\text{cyc}(\text{A})}}{\text{MF}-_{\text{cyc}(\text{A})}}\right)^2} \times \text{Signal}_{\text{Mediator}}$$

With:

- MF +/−: mean raw fluorescence signal of mediator positive or no mediator control samples, respectively.
- Cyc (A): all cycles before hot start.

- Cyc (B): cycle with UR signal maximum in mediator positive ±2 cycles.

- SD: standard deviation.

With fluorescence data presented in Fig. 7 for UR1 we receive $Signal_{Mediator} = 5.5$ (± 0.1).

In general, the direct incubation of a UR with the mediator in the presence of an active polymerase leads to the maximum fluorescence signal increase at the UR. With a $Signal_{Mediator} < 3.0$ in this characterization a low signal increase in MP PCR and thus a higher limit of detection must be expected. Low signal increase of the UR might be due to low quenching efficiencies. Therefore the fluorescence modifications should be considered as described in Subheading 3.5.2. Another reason for low signal increase might also be an ineffective mediator hybridization and extension, which can be improved by mediator design optimization (*see* Subheading 3.1.2).

4 Notes

1. *General troubleshooting*

   In case of low amplification efficiencies, resulting for example in high detection limits of the MP PCR, the same optimization steps as for other real-time PCR formats can help to improve results:

   - For primer- and probe sequence optimization we recommend the considerations published by J. SantaLucia [6].

   - Primer- and probe concentration optimization [17, 18, 26].

   - PCR reagent optimization like adaption of the Mg^{2+}-, polymerase-, or dNTP concentration or usage of additives such as DMSO [6, 17, 18].

2. *Temperature for fluorescence readout*

   The UR fluorescence signal is only efficiently quenched if the stem–loop structure is present as depicted in Fig. 6. To ensure no unspecific signal generation by temperature-mediated unfolding of the stem–loop structure, the fluorescence readout temperature must be accordingly adapted. In case of two-thermal step MP PCR protocols as presented in the models in this tutorial, readout is performed at the primer annealing and extension temperature, e.g., 58–63 °C. If three-thermal step MP PCRs are performed, the readout must be done at the primer annealing temperature and not at the ~72 °C extension temperature.

3. *Co-detection of analyte and IAC*

   In comparison to amplification reactions with one target as template more nonspecific side reactions have to be expected in case of duplex PCRs and thus a reduced reaction efficiency in comparison to single reactions. Before running the co-detection of two or more targets they should be tested as singleplex reactions—designed according to the recommendations in Subheading 3.2 and show acceptable performance.

4. *Fluorescence modification of UR*
For selection of the fluorophores or the fluorophore position in the stem, quenching effects of nucleobases must be considered [9]. Novel findings with UR2 revealed better results changing the guanosines adjacent to the fluorescein-modification at the thymidine to cytosines (the reverse complementary stem section must accordingly adapted).

Acknowledgments

We gratefully acknowledge funding by the German Research Foundation (DFG, contract number FKZ STE 1937/1-1). We further thank Sandra Cindric for support in the laboratory and Martin Trotter for proof reading.

References

1. Livak KJ, Flood SJ, Marmaro J et al (1995) Oligonucleotides with fluorescent dyes at opposite ends provide a quenched probe system useful for detecting PCR product and nucleic acid hybridization. PCR Methods Appl 4:357–362

2. Tyagi S, Kramer FR (1996) Molecular beacons: probes that fluoresce upon hybridization. Nat Biotechnol 14:303–308

3. Wittwer CT, Herrmann MG, Moss AA et al (1997) Continuous fluorescence monitoring of rapid cycle DNA amplification. Biotechniques 22:130–138

4. Gardner SN, Kuczmarski TA, Vitalis EA et al (2003) Limitations of TaqMan PCR for detecting divergent viral pathogens illustrated by hepatitis a, B, C, and E viruses and human immunodeficiency virus. J Clin Microbiol 41: 2417–2427

5. Lunge VR, Miller BJ, Livak KJ et al (2002) Factors affecting the performance of 5′ nuclease PCR assays for Listeria monocytogenes detection. J Microbiol Methods 51:361–368

6. SantaLucia J Jr (2007) Physical principles and visual-OMP software for optimal PCR design. In: Yuryev A (ed) Methods in Molecular Biology, vol 402. Humana Press, Totowa, pp 3–34

7. Livak KJ (1999) Allelic discrimination using fluorogenic probes and the 5′ nuclease assay. Genet Anal 14:43–149

8. Althaus CF, Gianella S, Rieder P et al (2010) Rational design of HIV-1 fluorescent hydrolysis probes considering phylogenetic variation and probe performance. J Virol Methods 165: 151–160

9. Marras SAE, Kramer FR, Tyagi S (2002) Efficiencies of fluorescence resonance energy transfer and contact-mediated quenching in oligonucleotide probes. Nucleic Acids Res 30:e122

10. Wong ML, Medrano JF (2005) Real-time PCR for MRNA quantitation. Biotechniques 39:75–85

11. Letertre C, Perelle S, Dilasser F et al (2003) Evaluation of the performance of LNA and MGB probes in 5′-nuclease PCR assays. Mol Cell Probes 17:307–311

12. Faltin B, Zengerle R, von Stetten F (2013) Current methods for fluorescence-based universal sequence-dependent detection of nucleic acids in homogenous assays and clinical applications. Clin Chem 59:1567–1582

13. Mokany E, Bone SM, Young PE et al (2010) MNAzymes, a versatile New class of nucleic acid enzymes that Can function as biosensors

and molecular switches. J Am Chem Soc 132:1051–1059

14. http://www.illumina.com/products/nupcr.ilmn. Accessed June 2013

15. http://www.seegene.com/en/research/core_040.php. Accessed June 2013

16. Faltin B, Wadle S, Roth G et al (2012) Mediator probe PCR: a novel approach for detection of real-time PCR based on label-free primary probes and standardized secondary universal fluorogenic reporters. Clin Chem 58: 1546–1556

17. King N (2010) Methods in Molecular Biology, 2nd edn, RT-PCR protocols. Humana, Totowa

18. Kennedy S, Oswald N (2011) PCR troubleshooting and optimization: the essential guide. Caister Academic Press, Poole

19. http://frodo.wi.mit.edu. Accessed June 2013

20. http://eu.idtdna.com/PrimerQuest/Home/Index

21. http://www.dnasoftware.com/product/visual-omp. Accessed June 2013

22. http://rna.tbi.univie.ac.at/cgi-bin/RNAfold.cgi. Accessed June 2013

23. Deprez RHL, Fijnvandraat AC, Ruijter JM et al (2002) Sensitivity and accuracy of quantitative real-time polymerase chain reaction using SYBR green I depends on cDNA synthesis conditions. Anal Biochem 307:63–69

24. http://blast.ncbi.nlm.nih.gov. Accessed June 2013

25. http://www.faculty.ucr.edu/~mmaduro/random.htm. Accessed June 2013

26. Gunson R, Gillespie G, Carman F (2003) Optimisation of PCR reactions using primer chessboarding. J Clin Virol 26:369–373

Chapter 7

Absolute Quantification of Viral DNA: The Quest for Perfection

Domenico Russo and Mauro Severo Malnati

Abstract

In spite of the impressive technical refinement of the PCR technology, new-generation real-time PCR assays still suffer from two major limitations: the impossibility to control both for PCR artifacts (with the important caveat of false-negative results) and for the efficiency of nucleic acid recovery during the preliminary extraction phase of DNA from the biological sample.

The calibrator technology developed at the Unit of Human Virology overcomes both of these limitations, leading to a substantially higher degree of accuracy and reproducibility in the quantification, which is especially useful for the measurement of pathogen loads in sequential samples and for the reliable detection of low-copy pathogens.

Key words Real-time PCR, Calibration, Absolute quantification, DNA viral load, HHV-6

1   Introduction

The increasing use of quantitative PCR-based assays in molecular diagnostics has prompted the development of new technologies, well suited for clinical routine, which combine the use of fluorogenic probes TaqMan, Fret, Beacons, Scorpions [1–9], with new types of instrumentation (ABI PRISM [7500, 7900], Light-cycler, I-cycler, Sentinel, etc.). These technologies allow for the simultaneous amplification and quantification of DNA in "real-time." By virtue of these new methods, which repeatedly measure target DNA during the exponential phase of PCR amplification [1], it is now possible to combine a high level of accuracy and reproducibility with an extreme sensitivity and detection range. Moreover, these technologies offer several advantages over traditional PCR methods:

1. Labor, costs, and sample handling time are greatly reduced because a single PCR run without further post-amplification steps is sufficient to accurately quantify target DNA.

Roberto Biassoni and Alessandro Raso (eds.), *Quantitative Real-Time PCR: Methods and Protocols*, Methods in Molecular Biology, vol. 1160, DOI 10.1007/978-1-4939-0733-5_7, © Springer Science+Business Media New York 2014

2. The absence of post-amplification manipulation steps greatly reduces the risk of inter-sample contamination and eliminates the need of radioactive or hazardous reagents.

3. By virtue of their high-throughput format, these systems are well suited for robotization and large epidemiological surveys.

Nevertheless, new-generation real-time PCR assays still suffer from two major limitations: the impossibility to control both for PCR artifacts, with the important caveat of false-negative results (Fig. 1), and for the efficiency of nucleic acid recovery during the preliminary extraction phase from the biological sample which causes a significant and unpredictable degree of variability (Table 1). Therefore, to fully exploit the power of the PCR technology, it is essential to introduce a tracer system that allows for the control and normalization of both steps of the analytical procedure: purification of nucleic acids followed by nucleic acid amplification and quantification. Our calibrator technology is based on the development of a "universal" synthetic DNA molecule with suitable base composition and sequence that guarantee the same kinetics of

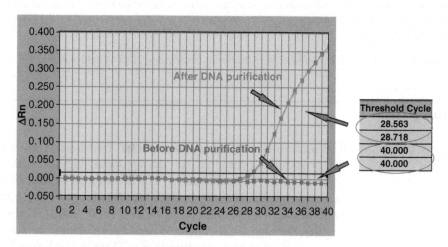

Fig. 1 Real-time PCR is affected by PCR inhibitors

Table 1
DNA recovery in different biological fluids

DNA input	Plasma	Serum	Liquor	Urine
250,000 *Geq	250,000	170,000	47,000	6,600
(%)	100	68	19	2.6
2,500,000 Geq	1,000,000	1,050,000	330,000	81,000
(%)	40	42	13	3.2

*Geq = HHV-6 Genome equivalent

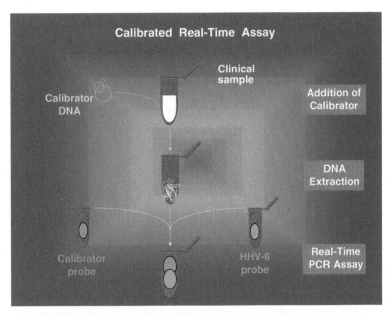

Fig. 2 Schematic representation of the calibrated real-time PCR technique

amplification as the target PCR amplicons [10–14]. The calibrator is spiked into the test sample before the DNA extraction step, co-purified along with the target DNA, and finally amplified (Fig. 2) using a dedicated set of primers and probe [10–14]. The latter is accurately chosen in order to avoid cross amplification with the target genome. The ratio between the amount of calibrator detected and the initial quantity spiked into the sample provides an accurate built-in index of the relative loss of DNA during the extraction phase, as well as of the efficiency of the amplification phase, which is particularly sensitive to the presence of Taq-polymerase inhibitors (Fig. 1). Noteworthy, the loss of the amplification signal of the calibrator molecule allows for the detection of false-negative results. Moreover, the addition of a known amount of calibrator molecule before any manipulation of the biological sample permits to measure the yield of DNA recovery for each sample, which varies significantly from sample to sample [10, 11, 15] and according to the source of the clinical specimens (Table 1). Thus, the calibrator controls each sample for the presence of PCR inhibitors, determines a cut-off value of sensitivity for negative samples, and normalizes positive samples for the efficiency of DNA recovery allowing higher accuracy and reproducibility of sample measurements as compared to the uncalibrated real-time PCR reaction (Table 2). The calibrator features combined with the real-time PCR technology allow to set up diagnostic assays perfectly suited for the clinical follow-up of infected patients [16–20], representing, in addition, valid tools to evaluate in vitro drug susceptibility of viral pathogens [21].

Table 2
Performance of the calibrated real-time HHV-6 assay

HHV-6 input[a] (copies/ml)	Mean of recovery (±SD)		Accuracy error (%)		Intersample CV[b] (%)	
	w/o calibration	With calibration	w/o calibration	With calibration	w/o calibration	With calibration
50	32 (±54)	45 (±25)	36.5	10	55	36
500	234 (±96)	367 (±118)	35.7	13.9	17.6	18.3
5,000	3,450 (±1,360)	5,150 (±310)	31	−3	36	19
500,000	269,132 (±54,375)	439,081 (±18,571)	42.1	6.4	37.5	9.5

[a]U1102 HHV-6 A DNA spiked in 1 ml aliquots of fetal bovine serum
[b]Values calculated using a set of ten replicate samples for each dilution point tested

2   Materials

Prepare all solutions using ultrapure sterile water and analytical grade reagents. Prepare and store all the reagents at room temperature unless otherwise indicated. Wear gloves to prepare all reagents.

2.1 Sample Preparation and DNA Extraction

1. Heat-inactivated fetal bovine serum (FBS): thaw FBS at 37 °C in water bath. When it is completely thawed increase the water temperature up 56 °C and leave FBS into the water for additional 45 min. Dispense heat-inactivated FBS in 50 ml falcon tube and centrifuge for 15 min at $1,600 \times g$. Remove the supernatant leaving pelleted material in the tube and filter (0.2 μm) in a sterile environment (i.e., under a laminar flow hood). Store FBS at −20 °C.

2. TES buffer: 10 mM Tris–HCl pH 8, 5 mM EDTA, and 0.5 % SDS. Mix 1 ml of Tris–Cl pH 8 1 M, 1 ml of EDTA 0.5 M, 5 ml of SDS 10 % solution, and 93 ml of water to obtain 100 ml of TES buffer.

3. Lysis Buffer: For each sample use 50 μl of TES buffer, 10 μl of proteinase K solution (20 mg/ml), 10 μl of calibrator at 10^3 copies/μl (see **Note 1**), and 330 μl of sterile water.

4. AE buffer: 5 mM Tris–Cl pH 8, 0.5 mM EDTA.

5. NaCl 5 M: Dissolve 292.2 g of NaCl in 800 ml of water. Once it is dissolved add water to reach the final volume of 1,000 ml.

6. Phenol/chloroform/isoamyl alcohol (PCIAA) 25/24/1 vol/vol: mix 25 ml Tris–HCl pH 8 equilibrated phenol with 24 ml of chloroform and 1 ml of isoamyl alcohol to obtain 50 ml PCIAA (see **Note 2**).

7. Chloroform/isoamyl alcohol (CIAA) 24/1 vol/vol mix (*see* **Note 3**).

8. Glycogen solution: dissolve 100 mg of glycogen in 10 ml of water (10 mg/ml solution).

9. Gloves.

2.2 Standard Curve Preparation

1. HHV-6 standard: Plasmid carrying the U67 133 bp fragment at the final concentration of 50 μg/ml (*see* **Note 1**).

2. TE buffer: 10 mM Tris–HCl pH 8, 5 mM EDTA. Mix 1 ml of Tris–Cl pH 8 1 M, 1 ml of EDTA 0.5 M, and 98 ml of water to obtain 100 ml of TE buffer.

3. Gloves.

2.3 Plate Setup

1. Real-time PCR mastermix (*see* **Note 4**).

2. HHV-6 primers and probe (Table 2).

3. Calibrator primers and probe (Table 2).

4. Gloves.

3 Methods

Carry out all procedures at room temperature (RT) unless otherwise specified.

3.1 Preparation of the Clinical Sample

1. Precool the centrifuge to the final temperature of 4 °C.

2. Transfer 1 ml of plasma or other biological fluids (serum, urine, cerebrospinal fluids, etc.) in a 2 ml eppendorf tube (*see* **Note 5**).

3. Centrifuge plasma or other biological fluids at $20,000 \times g$ for 1 h keeping the temperature at 4 °C.

4. Remove carefully the supernatant avoiding to touch the bottom of the tube. Leave 10–30 μl of the supernatant to be sure not to disturb the pelleted fraction (often invisible).

5. Add 100 μl of 0.2 μm filtered FBS.

6. Add 10–15 μl of calibrator (10^3 copies/μl) for each sample. The addition of a variable amount of calibrator is dependent by the final volume in which the purified DNA will be resuspended: typically 100 μl of final volume requires 10 μl of calibrator (1/10 for the final volume). Once the operator decide which is the volume needed to perform the real-time PCR analysis the calibrator can be added directly to the lysis buffer utilized to process the clinical sample.

7. Proceed immediately to DNA extraction. Since the calibrator is a double-stranded plasmid DNA the addition in the clinical sample need to be done jointly or followed immediately by the

addition of the lysis buffer that inactivates the DNAse present in the residual biological fluids and in the added FBS (*see* **Note 6**).

8. Using the same aliquot of calibrator used for sample spiking prepare a 1:10 dilution in TE buffer (typically 50 μl of calibrator in 450 μl of TE) and store both the undiluted and the 1/10 dilution at 4/−20 °C until their use as control of the calibrator input.

3.2 DNA Extraction

1. Add 400 μl of lysis buffer to the processed clinical sample, mix by vortexing for 20 s and incubate for 2 h at 56 °C.

2. Prepare a phenol/chloroform/isoamyl alcohol (PCIAA) 25/24/1 vol/vol mix.

3. Add 500 μl of PCIAA to each sample and vortex for 30 s.

4. Centrifuge at $8,000 \times g$ for 5 min at RT.

5. Prepare two 2 ml eppendorf tube containing 500 μl of chloroform/isoamyl alcohol 24/1 (CIAA) vol/vol mix (two eppendorf tube for each processed sample).

6. Carefully remove the aqueous phase from the spinned sample (upper part of the tube) avoiding to aspirate the material present at the PCIAA interface and add the aqueous phase to the first CIAA containing tube (*see* **Note 7**).

7. Centrifuge at $8,000 \times g$ for 5 min at RT.

8. Carefully remove the aqueous phase from the spinned sample (upper part of the tube) avoiding to aspirate the material present at the CIAA interface and add the aqueous phase to the second CIAA containing tube.

9. Centrifuge at $8,000 \times g$ for 5 min at RT.

10. Prepare a new 2 ml eppendorf tube containing 450 μl of absolute isopropanol, 50 μl of NaCl 5 M, and 5 μl of glycogen solution.

11. Repeat **step 8** adding the aqueous phase to the new tube.

12. Mix by inversion and leave samples to precipitate DNA at −20 °C for a minimum of 2 h (preferable O/N).

13. Precool the centrifuge to the final temperature of 4 °C.

14. Centrifuge at $8,000 \times g$ for 1 h at 4 °C.

15. Carefully remove the supernatant avoiding to touch the pelleted material (if visible) or the bottom of the tube.

16. Add 500 μl of 70 % Ethanol.

17. Centrifuge at $12,000 \times g$ for 20 min at 4 °C.

18. Remove completely the supernatant avoiding to touch with the tip the bottom of the tube.

19. Let air dry the pellet for 5 min under a sterile hood.

20. Resuspend DNA adding 100 µl of AE buffer and mixing for 30 min at 70 °C on a thermomixer.

21. Store DNA sample at 4 °C if you plan a real-time PCR run within a week otherwise at –20 °C for longer period of time.

3.3 Standard Curve Preparation

As a first step the quantification of the plasmidic DNA containing the HHV-6 target sequence need to be determined by UV-spectroscopy determining the 260/280 DNA absorbance ratio (optimal ratio = 1.9). For the HHV-6 plasmid or the calibrator containing plasmid the concentration of 50 µg/ml (0.05 µg/µl) correspond to $\approx 1 \times 10^{13}$ molecules/ml (1×10^9 copies/µl).

In general a clean bench (a flow laminar hood is preferred) need to be prepared washing it with denaturing agents such as NaClO and SDS or the commercially available preassembled solution DNA Zap before to proceed.

It is mandatory that this area must not be in the same room where clinical samples are processed or loaded in the PCR plate.

A dedicated set of calibrated micropipettes must be utilized and gloves need to be weared all along the entire procedure.

1. Prepare 12 eppendorf tubes (1.5–2 ml) containing 0.9 ml of TE buffer.

2. Warm-up the HHV-6 stock DNA (0.05 µg/µl = 10^9 copies/µl) for 5 min at 70 °C.

3. Cool on ice for 20 s.

4. Pipet 100 µl of the stock DNA into the first eppendorf tube emptying the tip once (without washing and resuspending) and avoiding to submerge the tip into the buffer closing immediately the HHV-6 DNA stock.

5. Mix by inversion (10–15 times) or by gentle vortexing.

6. Spin briefly to remove droplets from the cap of the tube.

7. Remove the HHV-6 DNA stock from the bench (*see* **Note 8**).

8. Repeat the complete procedure 11 times (last dilution = 10^{-2}copies/µl) (*see* **Note 8**).

Change gloves each time you prepare each dilution point (*see* **Note 9**).

3.4 Multiplex Real-Time PCR Plate Setup

The reaction mix and plate assembly need to be carried out in a clean and dedicated area. A small room equipped with a small –20 °C freezer, a small bench, and a closed cabinet in which to store all the dedicated micropipettes (one p20 and one p1000) and plasticware (microtips with filter, optical caps, and PCR plates) is the optimal solution. As an alternative a dedicated clean hood placed in a safe corner in one room where neither DNA extraction not sample DNA loading is carried out can be used.

Table 3
Primers and probes sequence

Real-time PCR system		Sequence	Probe labelling
HHV-6	Forward primer	5′-CAAGCCAAATTATCCAGAGCG-3′	
	Reverse primer	5′-CGCTAGGTTGAGGATGATCGA-3′	
	Probe	5′-CCCGAAGGAATAACGCTC-3′	FAM-MGB
Calibrator	Forward primer	5′-CCGGAAACCGAACATTACTGAA-3′	
	Reverse primer	5′-TTACGTGAGGATGATCGAGGC-3′	
	Probe	5′-ACGCCAACAGACCTAGCGA-3′	VIC-MGB

1. Thaw the real-time PCR mastermix (*see* **Note 4**).

2. Thaw both forward and reverse primers as well the probes for HHV-6 and calibrator (*see* **Note 10**).

3. In an eppendorf tube prepare the reaction mix accordingly to the concentrations reported in Table 3 calculating the number of wells employed in the experiment. Remember to take in account not only the wells employed for samples testing (both unknown and negative controls) but at least 15 wells for the Standard curve, 4 wells for controlling the calibrator input, and 6 wells of no template controls (*see* **Note 11**).

4. Mix the reaction mix by manual inversion (five times).

5. Fast spin ($6,000 \times g$ for 15 s) to remove droplets from the cap of the tube.

6. Dispense 15 μl in each well of the real-time PCR plate.

7. Cover the plate with an aluminum foil and transfer it in the loading hood.

3.5 Sample Loading

1. Transfer DNA samples, HHV-6 standard curve, calibrator input controls from the fridge/freezer to a thermomixer set to 56 °C and leave in gentle agitation for 10 min.

2. Centrifuge each sample at $6,000 \times g$ for 15 s to remove drops from the cap.

3. Add 10 μl of sample for each well of the preset PCR plate disposing samples in vertical lines (eight wells for line) and changing the tip each time you change sample (*see* **Note 12**). Load at least three wells for each sample.

4. Close each well with optical caps every time a line is filled.

3.6 Real-Time PCR Run Conditions

The described procedure has been optimized on a 7500 Fast Real-time PCR system. A different instrumentation might require a slight modification of the PCR run protocol.

1. Set the PCR run protocol starting with one cycle of DNA denaturation and enzyme activation lasting 15 min at 95 °C.

2. Forty cycles of annealing extension lasting 1 min at 60 °C intermingled by 40 cycles of denaturation at 95 °C for 15 s.

3.7 Calculation of the Recovery Rate and DNA Normalization

1. Measure the Calibration reference dilution samples appositely prepared during the extraction phase (both the 10^3 copies/μl and its 1:10 dilution point) of the biological sample (*see* **Note 13**).

2. Calculate recovery rate of the calibrator DNA (CI/CO) on positive samples and normalize the HHV-6 DNA load using the following formula:

$$VT / VR \times GER / VS \times CI / CO,$$

where VT indicates the total volume of the extracted material, VR the volume assayed in the PCR run, GER the number of HHV-6 genome equivalents measured in each reaction, VS the volume of the analytical sample expressed in milliliters, CI the calibrator input, and CO the calibrator output.

3. On PCR amplification negative samples the cutoff of sample sensitivity is calculated as follow: MGED/VS×CI/CO, where MGED indicates the minimal number of genome equivalent detectable, VS the volume of the biological sample express in milliliters, CI the calibrator input, and CO the calibrator output.

4   Notes

1. Both plasmids containing calibrator and HHV-6 U-67 target sequences can be purchased from Labospace Srl, Milan Italy.

2. The phenol/chloroform/isoamyl alcohol (PCIAA) 25/24/1 vol/vol mix can be purchased already assembled.

3. The chloroform/isoamyl alcohol (CIAA) 24:1 vol/vol mix can be purchased already assembled.

4. The described protocol has been set up using the Brilliant Multiplex QPCR Master Mix by STRATAGENE (Table 4). If other sources of real-time PCR mastermix are used set up experiments with both HHV-6 and Calibrator standard curves run alone and in combination need to be performed in order to ensure that both targets are coamplified with the same efficiency.

5. This step can be avoided when the extraction method used (i.e., Easy or Mini-mag system) can be directly applied to at least 1 ml of biological fluid.

6. Calculate the amount of Calibrator needed for each PCR run (typically 10^3 copies/μl, 10 μl for each sample) and add the entire amount directly in the batch of lysis buffer needed for

Table 4
Real-time PCR mix composition

			μl/well	μl/no. wells
Master mix			12.50	$n \times 12.50$
HHV-6	Forward primer	30 mM	0.25	$n \times 0.25$
	Reverse primer	30 mM	0.25	$n \times 0.25$
	Probe	15 mM	0.33	$n \times 0.25$
Calibrator	Forward primer	5 mM	0.25	$n \times 0.25$
	Reverse primer	5 mM	0.25	$n \times 0.25$
	Probe	2.5 mM	0.33	$n \times 0.25$
	Total volume		15.00	$n \times 15.00$

treating the entire set of samples. In this way you prevent the DNA degradation of the calibrator ensuring a higher reproducibility of your results handling a higher volume of calibrator with one pipette sampling.

7. The PCIAA purification step can be repeated at least a second time when the amount of material present at the interface is particularly abundant. In this case is mandatory repeat the CIAA step at least twice. Additional CIAA steps might be useful when the biological fluid processed is rich in fats.

8. Once you have prepared the first dilution (10^9 copies/μl) close the HHV-6 stock solution and place it in the –20 °C freezer in a separate box from where you are going to keep the point of the standard curve you plan to routinely use before starting to prepare the additional dilution points. Rapidly remove all the dilution points (as soon as you prepare the next dilution point) that are not commonly utilize in the PCR run. A standard curve is routinely used starting from 10^5 copies/μl to 10^{-1} copies/μl. Stock all the standard curve dilutions in small aliquots (30–50 μl each) in order to avoid utilizing several time the same aliquot. This procedure allow for a better interassay repeatability and reproducibility performance. All the dilutions with a concentration above 10^5 copies/μl can be stored in the same box with the stock solutions. The preparation of a 10^{-2} copies/μl dilution is required to correctly establish both the correctness of the dilution preparations and the absence of target DNA contamination. Typically this dilution point run in quintuplicates must yield the absence of amplification signal.

9. Pay extreme attention to not contaminate with the target DNA the set of dilution tubes employed for preparing the

standard curve. The glove that covers the hand used to open the cap of the eppendorf tube must be changed all the times.

10. The concentrations of calibrators primers (75 nM) and probe (50 nM) have been optimized with the HHV-6 assay to allow the distinction of the amplification signal of the two commonly used fluorochromes FITC (HHV-6 probe) and VIC (calibrator probe) avoiding at the same time emission interference due to the spectral overlap of the fluorochromes emission signal on a 7500 sequence detector system. The use of different fluorochromes and/or a different instrumentation might require a specific tune-up of primers and probe concentrations of the calibrator PCR.

11. Run duplicate points for each dilution of the standard curve utilized and a triplicate set for the last 10^{-1} dilution (input of 1 copy/reaction well). Regarding the calibrator once you have validated the reaction setup running at least three times a complete calibrator standard curve prepare aliquots of the 10^3 copies/reaction dilution (100–200 µl each). These aliquots will be utilized to spike the calibrator in the lysis buffer and as control of the calibrator input for determining the recovery ratio of each clinical sample. In this way, as long as you use the same batch of primers, probe, and mastermix utilized for validating the Calibrator curve you do not need to insert in the reaction plate also a standard curve for the calibrator. It will be sufficient to control that the Cq of the calibrator input and its 1/10 dilutions loaded in the reaction plate are consistent with the ones obtained during the validation process.

12. We suggest to dispense samples following the vertical rows of the plate and to run triplicates for each clinical samples. Optimize sample loading trying to fill each line avoiding to load wells containing the same clinical sample in different rows. In this way it will be possible to close each row separately as soon as it is filled limiting the potential cross contamination between samples. In addition it is useful to insert six to eight well of no template controls as well as three extraction negative controls samples intermingled to the clinical samples in different position of the plate (bottom, center, top). It is important remember to load the standard curve dilutions starting from the lower concentrations (from 10^{-1} up) once that all the clinical samples are loaded and cap sealed.

13. The choice to utilize 10^4 copies of calibrator in each biological sample is determined by the minimal concentration of calibrator measured with an optimal degree of repeatability and reproducibility (typically a $CV < 0.1$ quantitation cycle) given that routinarily 1/10 (10^3 copies) of the volume of the extracted material is loaded in each PCR well.

References

1. Heid CA, Stevens J, Livak KJ et al (1996) Real-time quantitative PCR. Genome Res 6:986–994

2. Kalinina O, Lebedeva I, Brown J et al (1997) Nanoliter scale PCR with TaqMan detection. Nucleic Acids Res 25:1999–2004

3. Whitcombe D, Brownie J, Gillard HL et al (1998) A homogeneous fluorescence assay for PCR amplicons: its application to real-time, single-tube genotyping. Clin Chem 44:918–923

4. Solinas A, Brown LJ, McKeen C et al (2001) Duplex Scorpion primers in SNP analysis and FRET applications. Nucleic Acids Res 29:E96

5. Gelsthorpe AR, Wells RS, Lowe AP et al (1999) High-throughput class I HLA genotyping using fluorescence resonance energy transfer (FRET) probes and sequence-specific primer-polymerase chain reaction (SSP-PCR). Tissue Antigens 54:603–614

6. Chen W, Martinez G, Mulchandani A (2000) Molecular beacons: a real-time polymerase chain reaction assay for detecting Salmonella. Anal Biochem 280:166–172

7. Chen X, Livak KJ, Kwok PY (1998) A homogeneous, ligase-mediated DNA diagnostic test. Genome Res 8:549–556

8. Carters R, Ferguson J, Gaut R et al (2008) Design and use of scorpions fluorescent signaling molecules. Methods Mol Biol 429:99–115

9. Ortiz E, Estrada G, Lizardi PM (1998) PNA molecular beacons for rapid detection of PCR amplicons. Mol Cell Probes 12:219–226

10. Broccolo F, Locatelli G, Sarmati L et al (2002) Calibrated real-time PCR assay for quantitation of human herpesvirus 8 DNA in biological fluids. J Clin Microbiol 40:4652–4658

11. Broccolo F, Scarpellini P, Locatelli G et al (2003) Rapid diagnosis of mycobacterial infections and quantitation of *Mycobacterium tuberculosis* load using two real-time calibrated PCR assays. J Clin Microbiol 41:4565–4572

12. Flamand L, Gravel A, Boutolleau D et al (2008) Multicenter comparison of PCR assays for detection of human herpesvirus 6 DNA in serum. J Clin Microbiol 46:2700–2706

13. Gatto F, Cassina G, Broccolo F et al (2011) A multiplex calibrated real-time PCR assay for quantitation of DNA of EBV-1 and 2. J Virol Methods 178:98–105

14. Cassina G, Russo D, De Battista D et al (2013) Calibrated real-time polymerase chain reaction for specific quantitation of HHV-6A and HHV-6B in clinical samples. J Virol Methods 189:172–179

15. Tedeschi R, Enbom M, Bidoli E et al (2001) Viral load of human herpesvirus 8 in peripheral blood of human immunodeficiency virus infected patients with Kaposi's sarcoma. J Clin Microbiol 39:4269–4273

16. Malnati MS, Broccolo F, Nozza S et al (2002) Retrospective analysis of HHV-8 viremia and cellular viral load in HIV-seropositive patients receiving interleukin 2 in combination with antiretroviral therapy. Blood 100:1575–1578

17. Dagna L, Broccolo F, Paties CT et al (2005) A relapsing inflammatory syndrome and active human herpesvirus 8 infection. N Engl J Med 353:156–163

18. Broccolo F, Drago F, Careddu AM et al (2005) Additional evidence that pityriasis rosea is associated with reactivation of HHV-6 and -7. J Invest Dermatol 124:1234–1240

19. Drago F, Broccolo F, Zaccaria E et al (2008) Pregnancy outcome in patients with pityriasis rosea. J Am Acad Dermatol 58:S78–S83

20. Broccolo F, Drago F, Paolino S et al (2009) Reactivation of human herpesvirus 6 (HHV-6) infection in patients with connective tissue disorders. J Clin Virol 46:43–46

21. Broccolo F, Lusso P, Malnati M (2013) Calibration technologies for correct determination of Epstein-Barr virus, human herpesvirus 6 (HHV-6), and HHV-8 antiviral drug susceptibilities by use of real-time PCR-based assays. J Clin Microbiol 51:2013

Chapter 8

A Multiplex Real-Time PCR-Platform Integrated into Automated Extraction Method for the Rapid Detection and Measurement of Oncogenic HPV Type-Specific Viral DNA Load from Cervical Samples

Francesco Broccolo

Abstract

The persistent infection with most frequent high-risk (HR)-HPV types (HPV-16, -18, -31, -33, -45, -52, and -58) is considered to be the true precursor of neoplastic progression. HR-HPV detection and genotyping is the most effective and accurate approach in screening of the early cervical lesions and cervical cancer, although also the HR-HPV DNA load is considered an ancillary marker for persistent HPV infection.

Here, it is described an in-house multiplex quantitative real-time PCR (qPCR)-based typing system for the rapid detection and quantitation of the most common HR-HPV genotypes from cervical cytology screening tests.

First, a separate qPCR assay to quantify a single-copy gene is recommended prior to screening (pre-screening assay) to verify the adequate cellularity of the sample and the quality of DNA extracted and to normalize the HPV copy number per genomic DNA equivalent in the sample. Subsequently, to minimize the number of reactions, two multiplex qPCR assays (first line screening) are performed to detect and quantify HPV-16, -18, -31, -33, -45, -52, and -58 (HPV-18 and -45 are measured together by single-fluorophore). In addition, a multiplex qPCR assay specific for HPV-18 and HPV-45 is also available to type precisely the samples found to be positive for one of the two strains. Finally, two nucleic acid extraction methods are proposed by using a 96-well plate format: one manual method (supported by centrifuge or by vacuum) and one automated method integrated into a robotic liquid handler workstation to minimize material and hands-on time.

In conclusion, this system provides a reliable high-throughput method for the rapid detection and quantitation of HR-HPV DNA load in cervical samples.

Key words HPV, High-risk HPV, HPV genotyping, HPV genotypes

1 Introduction

It has been established that infection by oncogenic human papillomavirus (HPV) is a necessary condition for cervical carcinogenesis [1–5]. The most frequent high-risk HPV types are HPV-16, -18, -31, -33, -45, -52, and -58 [6]. Persistent infection is

Roberto Biassoni and Alessandro Raso (eds.), *Quantitative Real-Time PCR: Methods and Protocols*, Methods in Molecular Biology, vol. 1160, DOI 10.1007/978-1-4939-0733-5_8, © Springer Science+Business Media New York 2014

considered to be the true precursor of neoplastic progression [7, 8]. Currently, however, HPV infections are monitored primarily by qualitative HPV DNA detection assays which are often not type specific and therefore in the clinical management of the patients do not distinguish between persistent and transient infections, the latter being extremely frequent in sexually active women [1].

Although HPV genotyping is not required for routine HPV analysis, genotyping tests for HPV-16 and HPV-18 have been approved and recommended as an option for specific clinical scenarios to guide triage to colposcopy because they imply a higher risk for progression of dysplasia [9, 10]. Hybrid Capture 2 (HC2; Qiagen Gaithersburg, Inc., Gaithersburg, MD, USA) is the main commercially available HPV test to screen for HPV infection [6]. In addition, most studies report also that high-risk (HR-HPV) DNA load is an ancillary marker for persistent HPV infection [11–16]. However, only the normalized quantitative measurement assay such as quantitative real-time PCR (qPCR) allows to reliably determine HPV DNA levels: in fact, the normalization of the viral load for a number of cells is critical to verify the adequate cellularity of the sample, to control the recovery and the quality of DNA extracted. Moreover, the cumulative HPV viral load determined by HC2 represents the sum of multiple infections without the possibility to determine the real contribution of each HPV-genotype replication [17].

Finally, two nucleic acid extraction methods are proposed by using a 96-well plate format: one manual method (supported by centrifuge or by vacuum) and one automated method integrated into a robotic liquid handler workstation to minimize material and hands-on time.

Here it is described an in-house multiplex qPCR-based typing system for the rapid detection of the most common HR-HPV types providing a reliable and high-throughput method to determine the normalized HPV type-specific viral DNA load from cervical samples.

2   Materials

2.1 Sample Preparation from Cervical Samples

Cervical cytological materials are scraped from the endocervix using a rotary motion with a Cytobrush (Digene Cervical sampler, Digene Corp., Gaithersburg, MD, USA) and then collected in ThinPrep PreservCyt (Hologic) liquid media.

2.2 DNA Extraction by Nucleospin 96 Blood Kit

DNA extraction from cervical specimens is carried out by centrifugation or by vacuum manifolds utilizing a commercial kit based on binding of nucleic acid to silica membrane (NucleoSpin® 96 Blood kit (Macherey–Nagel) or similar membranes from other vendors).

All reagents for acid nucleic extraction are already included in the kit.

Before starting the DNA extraction to prepare Buffer B5 and Proteinase K solution according to the manufacturer's instructions:

1. Add the indicated volume of 96–100 % ethanol to the Buffer B5 Concentrate.

2. Store Wash Buffer B5 at room temperature (20–25 °C) for up to 1 year.

3. Add the indicated volume of Proteinase Buffer to lyophilized Proteinase K (this solution is stable at –20 °C for 6 months).

Instruments for manual DNA extraction by NucleoSpin® 96 Blood kit:

4. Microtiterplate centrifuge able to accommodate NucleoSpin® 96 Blood QuickPure Binding Strips/Plates stacked on an MN Square-well Block or Rack of Tube Strips which reaches accelerations of $5,600 \times g$ is required (e.g., Hermle Z513, Qiagen/Sigma 4-15c, Jouan KR4i, Kendro-Heraeus Multifuge 3/3-R, Highplate™, Beckman Coulter, Allegra R25).

5. NucleoVac 96 Vacuum Manifold (or other suitable vacuum manifolds).

The complete procedure under vacuum can be easily adapted to work station of a robotic liquid handler. Instruments for automate DNA extraction by NucleoSpin® 96 Blood kit:

6. Tecan Freedom Nucleic Acid Sample Preparation Workstation equipped with a Te-VacS vacuum module.

7. Biomeck 2000 (Beckman).

8. Microlab Star.

9. ®The CyBi®-RoboSpense.

2.3 Preparation of HR-HPV Plasmids

1. PCR products obtained using primers as reported (Table 1) are cloned into the pCRII plasmid by using the TOPO-TA cloning kit (Invitrogen Corp., San Diego, Calif.) according to the manufacturer's instructions.

2. Taq polymerase.

3. LB plates containing 50 μg/ml ampicillin or 50 μg/ml kanamycin.

4. 40 mg/ml X-gaL in dimethylformamide (DMF).

5. 100 mM IPTG in water.

6. 42 °C Water bath.

7. 37 °C Shaking and non-shaking incubator.

8. General microbiological supplies (e.g., plates, spreaders).

9. Plasmid Maxiprep reagents (i.e., Qiagen, Life Technologies or other vendors).

10. The subcloned cDNA of each construct is then sequenced.

2.4 Real-Time
PCR Assays

1. For all single-tube (multiplex) amplification reactions is recommended the use of master mixes specifically optimized to amplify more than one target of interest (up to four targets in one reaction) (i.e., Brilliant® Multiplex QPCR Master Mix (Agilent) or other similar reagent). A passive reference dye is included in this kit and may be added to compensate for non-PCR-related variations in fluorescence between wells (caused by slight volume differences in reaction tubes) (*see* **Note 1**).

2. The sequences of the primers for the amplification and probes for the detection are shown (Table 1).

   Two types of fluorogenic hybridization probes are utilized:

   (a) Single-labeled probes (in the assays to detect HPV-16, -18, -45, -33/52/58 and CCR5 gene).

   (b) Dual-labeled probes (in the assay to detect HPV-18/45 and 31).

Table 1
Primers and fluorescently labeled hydrolysis hybridization probes

Genotype (target sequence)	Sequence (5' to 3')	Labels (5', 3') fluorophore	Screening
CCR5 (chemokine receptor 5)	F: ATGATTCCTGGGAGAGACGC 3' R: AGCCAGGACGGTCACCTT P: AACACAGCCACCACCCAAGTGATCA	FAM, NFQ	Prescreening (reaction 1)
HPV-16 (E1 ORF)	F: CGAAAGTATTTGGGTAGTCCACTTA R: CAGCTCTACTTTGTTTTTCTATACATATGG P: AGTGAATGTGTAGACAATAA	VIC, NFQ	First line (reaction 1)
HPV-18/45 (E1 ORF)	F: TTTGAAAGGACATGGTCCAGAT R: CGTTCCGAAAGGGTTTCC P: AGATTTGCACGAGGAAGAGGAAGATGC	FAM, TAMRA	First line (reaction 1)
HPV-31 (E2 ORF)	F: CCACCACATCGAATTCCAA R: CGCCGCACACCTTCAC P: CCTGCGCCTTGGGCACC	VIC, TAMRA	First line (reaction 2)
HPV-33/52/58 (E1 ORF)	F: GGACGTGGTGCAAATTAGATTT R: TTTTCTCCTGCACTGCATTTAAAC P: AGGAAGAGGACAAGGAA	FAM, NFQ	First line (reaction 2)
HPV-18 E6 ORF)	F: TTTTGCTGTGCAACCGATT R: AGTGCCAGCGTACTGTATTGTG P: CGGTTGCCTTTGGCTT	FAM, NFQ	Second line (reaction 1)
HPV-45 (E6 ORF)	F: CAGTACCGAGGGCAGTGTAA R: CGTCTGCGAAGTCTTTCTTG P: ACATGTTGTGACCAGGC	VIC, NFQ	Second line (reaction 1)

All probes are labeled with 5' FAM (6-carboxyfuorescein) and 3' TAMRA (6-carboxytetramethylrhodamine)
MGB probe minor groove binder, *NFQ* non-fluorescent quencher, *F* forward primer, *R* reverse primer, *P* probe

The single-labeled probes have a minor groove binder (MGB) and non-fluorescent quencher (NFQ) at the 3′-end of the DNA sequence, whereas dual-labeled probes have a fluorescent quencher (TAMRA) at the 3′-end of the DNA sequence. HVP-18 and -33/52/58 probes are labeled to the 5′-end of the DNA sequence with 5′ fluorophore (FAM), while HPV-16, -31, and -45 with 5′ fluorophore (VIC).

3. Spectrofluorometric thermal cycler (ABI Prism 7000, 7700, 7900HT Applied Biosystems part of Life Technologies, Mx4000 and Mx3000 Stragene/Agilent or similar instruments by other vendors).

3   Methods

3.1   Cell Pellet Preparation from Cervical Samples

Cells are collected from the remnants of the ThinPrep cytology samples by centrifugation using the following procedure:

1. Decant Thin-prep cytology remnants into 50 ml tube.

2. Fill up to 45 ml with 1× PBS buffer. Invert several times.

3. Centrifuge at $1,500 \times g$ for 10 min to collect the cytological remnants to the bottom of 50 ml tube.

4. Decant (or vacuum-suck) the supernatant gently and carefully without disturbing the cell pellet.

5. Add 20 ml of 1× PBS and suspend the cells gently (no vortex).

6. Centrifuge at $1,500 \times g$ for 10 min to collect cells.

7. Decant (or vacuum-suck) the supernatant gently and carefully without disturbing the cell pellet.

8. Suspend the cells with the remaining 1× PBS inside the tube using 1 ml pipette tip and pipetteman.

9. Transfer the suspended cells to 1.5/2.0 ml flip-cap tubes.

10. Centrifuge at $13,000 \times g$ for 1 min and remove supernatant using pipette tips.

11. Suspend the cellular pellet with 1 ml of 1× PBS.

DNA extraction from cervical specimens is carried out by centrifugation or by vacuum manifolds utilizing NucleoSpin® 96 Blood kit (Macherey–Nagel).

3.2   Nucleic Acid Extraction: Centrifuge Processing

1. Pipette 25 μl Proteinase K and 200 μl of cells (obtained from Thinprep) into an MN Square-well Block.

2. Add 200 μl Buffer BQ1 to each sample. Cover MN Square-well Block with Self-adhering Foil.

3. Incubate MN Square-well Block at ambient temperature (18–25 °C) for 10 min on a shaker at high shaking speed (*see* **Note 2**).

4. Remove Self-adhering Foil and add 200 μl ethanol (96–100 %) to each well of the MN Square-well Block.

5. Mix the lysate by pipetting up and down three times.

6. Place a NucleoSpin® Blood QuickPure Binding Plate onto an empty MN Square-well Block.

7. Transfer lysates into the NucleoSpin® Blood QuickPure Binding Plate. Cover plate with a new Self-adhering Foil.

8. Centrifuge 3 min at $5,600 \times g$. If the samples are not drawn through the matrix completely, repeat the centrifugation step. Discard flow-through, clean, and sterilize MN Square-well for a 10 min period. It is not necessary to discard flow-through.

9. Remove Self-adhering Foil and add 500 μl Buffer BQ2 to each well of the NucleoSpin® Blood QuickPure Binding Plate. Cover it with a new Self-adhering Foil.

10. Centrifuge 5 min at $5,600 \times g$.

11. Place NucleoSpin® Blood QuickPure Binding Plate onto a Round-well Block. Remove Self-adhering Foil and add 100 μl elution Buffer BE to each well. Cover it with a new Self-adhering Foil.

12. Centrifuge 2 min at $5,600 \times g$. Seal Round-well Block (Low) with Self-adhering Foil for DNA storage. If elution in small volume tubes is desired, place a 96 PCR plate (not supplied) on top of the Rack of Tube Strips and elute into the PCR plate.

3.3 Nucleic Acid Extraction: Vacuum Processing

1. Transfer 200 μl of cells (obtained from Thinprep) into the wells of a Lysis Block. Place the plate on the vacuum manifolds or on the vacuum manifold integrated into workstation.

2. Add 100 μl of a premix of buffer BQ1 and proteinase K (75 μl buffer BQ1, 25 μl proteinase K) to each sample, mix by pipetting up and down for three times and shakes for 10 min.

3. Add 400 μl of a premix of ethanol and buffer BQ1 (200 μl ethanol, 200 μl buffer BQ1) and mix by pipetting up and down three times and shakes for 30 s.

4. Transfer the mixture to the NucleoSpin® Blood Binding Strips/Plate.

5. Overlaid the lysate with 150 μl buffer B5.

6. Apply vacuum 800 mbar for 5 min.

7. Wash three times the silica membrane by adding 600 μl of wash buffer BW and 2× 900 μl wash buffer B5. For each wash step, apply vacuum for 3 min at 800 mbar.

8. Remove the MN Wash Plate from the vacuum manifold with the robotic microplate handler disassembling and reassembling the vacuum manifold.

9. Dry the membrane using the automated column drying software mode with full vacuum for a minimum of 10 min at 600 mbar.

10. Elute purify DNA by adding 100 μl of pre-warmed Elution Buffer BE. Repeat once elution step. Apply Vacuum for 1 min at 600 mbar for both steps.

3.4 Preparation of Plasmid DNA Standards

1. Plasmids containing HPV-16, -18/45, -31, -33/52/58, -18, -45 sequences are prepared by cloning the amplified products (using clinical samples previously tested) obtained using the same forward and reverse primers used for the qPCR assays (Table 1).

2. Use primer concentrations and PCR cycling profile as shown (Tables 2 and 3).

3. Then PCR products are cloned into the pCRII plasmid by using the TOPO-TA cloning kit (Invitrogen Corp., San Diego, Calif.) according to the manufacturer's instructions.

4. The subcloned HPVs DNA constructs are then purified using Plasmid Maxiprep reagents (i.e., Qiagen, Life Technologies or other vendors) and then sequenced.

5. In order to generate reference curves for the determination of HPV-16, -18, -45, -18/45, -31, -33/52/58 copy numbers, plasmids are quantified by UV spectroscopy.

3.5 Real-Time PCR Assays

The qPCR assays for the HR-HPV flow chart is showed in Fig. 1 and include:

1. One reaction, to measure the amount of single-copy CCR5 gene (STEP 1 prescreening) (*see* **Note 3**); Two reactions, both in single-tube (multiplex): one to detect HPV-16, and -18/45 (with two different fluorophores) and one to detect HPV-31

Table 2
The PCR cycling profile

Cycles	Duration of cycle	Temperature (°C)
1	15 min	95
40	20 s	95
	30 s	60
	30 s	72
1	10 min	72

Table 3
Protocols for qPCR assays

Components	Volume (μl)	Concentration in final reaction
A. qPCR Assay for the measurement of CCR5 DNA load (*see* **Note 5**)		
2× Brilliant qPCR master mix	12.5	1×
10 μM Primer F_1	0.75	300 nM
10 μM Primer R_1	0.75	300 nM
5 μM $Probe_1$	1	200 nM
B. Multiplex qPCR assays for the detection HR-HPV DNA (*see* **Note 6**)		
2× Brilliant multiplex qPCR master mix	12.5	1×
5 μM Primer F_1	0.375	75 nM
5 μM Primer R_1	0.375	75 nM
5 μM Primer F_2	0.375	75 nM
5 μM Primer R_2	0.375	75 nM
2.5 μM $Probe_1$	0.5	50 nM
2.5 μM $Probe_2$	0.5	50 nM

The subscripts 1 and 2 indicate two reactions amplified in a single tube (i.e., HPV-16 and HPV-18/45, HPV-31 and HPV-33/52/58, or HPV-18 and HPV-45)

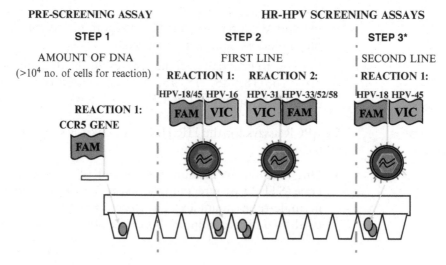

Fig. 1 Flow chart of qPCR assays for the HR-HPV screening

and -33/52/58 (with two different fluorophores) (STEP 2, screening first line) (*see* **Note 4**); One reaction in single-tube (multiplex) specific for HPV-18 and -45 disposable to the samples which are found to be positive for HPV-18 and/or -45 (*see* **Note 4**).

2. Prepare the experimental reaction by adding all the components (Table 3).

3. Gently mix the reactions without creating bubbles (do not vortex).

4. Add 10 µl of experimental DNA, or plasmid DNA or H_2O DNAse free (as no template control) to each experimental reaction.

5. Gently mix the reactions without creating bubbles (do not vortex).

6. Centrifuge the reactions briefly.

7. All reactions have been optimized to obtain the best amplification kinetics under the same cycling conditions (Table 2).

4 Notes

1. Although addition of the reference dye is optional when using the Mx4000, Mx3000P, or Mx3005P system, with other instruments (including the ABI 7900HT and ABI PRISM 7700) the use of the reference dye may be required for optimal results. Prepare fresh dilutions of the reference dye prior to setting up the reactions, and keep all tubes containing the reference dye protected from light as much as possible. Make initial dilutions of the reference dye using nuclease-free PCR-grade H_2O. If using the ABI PRISM 7700 instrument, use the reference dye at a final concentration of 300 nM (dilute the dye solution provided 1:50). If using the Stratagene Mx3000P, Mx3005P, or Mx4000 instrument, use the reference dye at a final concentration of 30 nM (dilute the dye solution provided 1:500).

2. If no shaker is available, pipette the lysate up and down three to five times to ensure thorough mixing of the solution. The lysate will turn brownish during incubation with Buffer BQ1 and Proteinase K. Increase incubation time with Proteinase K (up to 30 min) if processing cell pellet with more than 5×10^6 cells.

3. Quantitation of CCR5 copy number is performed separately from the HR-HPV qPCR assays since it cannot be standardized in a multiplex format. Details are described in a previous publication [18]. By CCR5 single-copy gene assay it is possible:

   (a) To verify the adequate cellularity of the sample (*see* **Note 7**).

   (b) To verify the quality of DNA extracted (*see* **Note 8**).

   (c) To normalize the HPV copy number per genomic DNA equivalent in the sample (*see* **Note 9**).

4. In these assays HPV-18/45 (HPV-18 and -45) as well as HPV-33/52/58 (HPV-33 and -52 and -58) are detected together by using single probes. Details are described in a previous publication [19, 20].

5. For CCR5 assay target-specific primers and probes are used at the final concentrations of 300 and 200 nM.

6. In multiplex assays, target-specific primers and probes are recommended, respectively, at the final concentrations of 75 and 50 nM to decrease the effect of cross talk due to emission spectrum overlap of the two dyes (FAM/VIC). The effect of cross talk is showed only when the difference between targets is higher of 2 logs.

7. Cervical samples differ widely in the amount of DNA present. DNA from cervical samples is considered suitable for HPV viral load determination if the human CCR5 copy number for reaction was higher than 2×10^4 (corresponding to 10^4 cells for reaction). Conversely, when the concentration of human genomic DNA is high (>1 μg) the sample can be diluted in order to have a final concentration not higher than (equal or lower than) 1 μg/reaction. In fact, concentration of human genomic DNA high (>1 μg) affects the sensitivity of HR-HPV assay. To resolve this problem each annealing/extension step for all cycles can be increased by 5 s in a stepwise fashion.

8. The presence of inhibitors in the DNA extracted is detected by kinetics of the amplification plots.

9. Normalization of HPV type-specific viral load is calculated as:

$$VL = \frac{Cn_{HPV}}{\left(Cn_{CCR5} / 2\right)} \times 10^4 \, \text{cells,}$$

where VL is the number of HPV genomes per 10^4 cells (corresponding to 2×10^4 CCR5 copies), Cn_{HPV} is the number of HPV genomes, and $Cn_{CCR5}/2$ is the number of cells.

References

1. Zur Hausen H (2002) Papillomaviruses and cancer: from basic studies to clinical application. Nat Rev Cancer 2:342–350

2. Ho GY, Bierman R, Beardsley L et al (1998) Natural history of cervicovaginal papillomavirus infection in young women. N Engl J Med 338:423–428

3. Woodman CBJ, Collins SI, Young LS (2007) The natural history of cervical HPV infection: unresolved issues. Nat Rev Cancer 7:11–22

4. Bosch FX, Muñoz N (2002) The viral etiology of cervical cancer. Virus Res 89:183–190

5. Bosch FX, Lorincz A, Muñoz N et al (2002) The causal relation between human papillomavirus and cervical cancer. J Clin Pathol 55:244–265

6. Muñoz N, Bosch FX, De Sanjose S et al (2003) Epidemiologic classification of human papillomavirus types associated with cervical cancer. N Engl J Med 348:518–527

7. Wallin KL, Wiklund F, Angström T et al (1999) Type-specific persistence of human papillomavirus DNA before the development of invasive cervical cancer. N Engl J Med 341:1633–1638

8. Kjaer SK, van den Brule AJ, Paull G et al (2002) Type specific persistence of high risk human papillomavirus (HPV) as indicator of high grade cervical squamous intraepithelial lesions in young women: population based prospective follow up study. BMJ 325:572

9. Massad LS, Einstein MH, Huh WK et al (2013) 2012 Updated consensus guidelines for the management of abnormal cervical cancer screening tests and cancer precursors. 2012 ASCCP consensus guidelines conference. Obstet Gynecol 121:829–846

10. Massad LS, Einstein MH, Huh WK et al (2013) 2012 Updated consensus guidelines for the management of abnormal cervical cancer screening tests and cancer precursors. 2012 ASCCP consensus guidelines conference. J Low Genit Tract Dis 17:S1–S27

11. Dalstein V, Riethmuller D, Pretet JL et al (2003) Persistence and load of high-risk HPV are predictors for development of high-grade cervical lesions: a longitudinal French cohort study. Int J Cancer 106:396–403

12. Flores R, Papenfuss M, Klimecki WT et al (2006) Cross-sectional analysis of oncogenic HPV viral load and cervical intraepithelial neoplasia. Int J Cancer 118:1187–1193

13. Gravitt PE, Burk RD, Lorincz A et al (2003) A comparison between real-time polymerase chain reaction and hybrid capture 2 for human papillomavirus DNA quantitation. Cancer Epidemiol Biomarkers Prev 12:477–484

14. Josefsson AM, Magnusson PK, Ylitalo N et al (2000) Viral load of human papilloma virus 16 as a determinant for development of cervical carcinoma in situ: a nested case–control study. Lancet 355:2189–2193

15. Lorincz AT, Castle PE, Sherman ME et al (2002) Viral load of human papillomavirus and risk of CIN3 or cervical cancer. Lancet 360:228–229

16. Moberg M, Gustavsson I, Wilander E et al (2005) High viral loads of human papillomavirus predict risk of invasive cervical carcinoma. Br J Cancer 92:891–894

17. Broccolo F, Fusetti L, Ceccherini-Nelli L (2012) Is it true that pre-conization high-risk HPV DNA load is a significant factor of persistence of HPV infection after conization? J Clin Virol 55:377–378

18. Malnati MS, Scarlatti G, Gatto F et al (2008) A universal real-time PCR assay for the quantification of group-M HIV-1 proviral load. Nat Protoc 3:1240–1248

19. Broccolo F, Cocuzza CE (2008) Automated extraction and quantitation of oncogenic HPV genotypes from cervical samples by a real-time PCR-based system. J Virol Methods 148:48–57

20. Broccolo F, Chiari S, Piana A et al (2009) Prevalence and viral load of oncogenic human papillomavirus types associated with cervical carcinoma in a population of North Italy. J Med Virol 81:278–287

Chapter 9

Real-Time PCR Detection of *Mycoplasma pneumoniae* in the Diagnosis of Community-Acquired Pneumonia

Eddi Di Marco

Abstract

Polymerase chain reaction is a useful technique in microbial diagnostics to detect and quantify DNA or RNA of low abundance.

Bacterial and viral nucleic acid can be amplified by PCR upon clinical sample extraction using specific primers for classical qualitative PCR and primers and probes for real-time PCR.

Here we describe the Scorpion-probe real-time PCR-based assay that offers thermodynamic advantages due to its kinetic reaction and provides faster performances compared to a classical double-labeled probe-based assays.

Key words *Mycoplasma pneumoniae*, Community-acquired pneumonia, Quantitative PCR, Scorpion probe

1 Introduction

Mycoplasma pneumoniae is a common cause of upper respiratory tract infection and is one of the etiological agents of community-acquired pneumoniae (CAP) [1].

The direct determination of *Mycoplasma pneumoniae* DNA using real-time PCR in clinical specimens allows an efficient detection of this etiological agent in all the phases of the infection, avoiding the false-negative serological assay responses during the first 1–2 weeks after the primary infection [2]. In addition the qPCR assay on clinical samples performed better than conventional qualitative PCR in terms of sensitivity and specificity, since it allowed the detection of the specific pathogen in some specimens where the classic qualitative PCR failed.

The use of real-time PCR in microbial molecular diagnostics can be clinically relevant also for the short-time results compared to traditional assays [3–5].

Different chemistries are employed to monitor the fluorescence emitted during the reaction as a function of amplicon production

Roberto Biassoni and Alessandro Raso (eds.), *Quantitative Real-Time PCR: Methods and Protocols*, Methods in Molecular Biology, vol. 1160, DOI 10.1007/978-1-4939-0733-5_9, © Springer Science+Business Media New York 2014

at each PCR cycle [6]. Real-time PCR using Scorpion unimolecular probe gives several important advantages, chief of which is faster reaction kinetics due to its intramolecular probing mechanism that ensures a near proximity between probe and target DNA [7]. In addition Scorpion probes do not need any fluorochrome enzymatic cleavage that occurred during the double-labeled probe-based assays. This allows a rapid-cycling PCR that reduces the overall procedure time by more than 30 %, ideal for its use in hospital settings.

2   Materials

2.1   Specimen Collection, Storage, Transport

1. Nasopharyngeal aspirates, nasopharyngeal swabs, and bronchoalveolar lavage specimens are routinely examined for *Mycoplasma pneumoniae* infection (*see* **Note 1**). All samples collected have to be treated as potentially infectious material.

2. The swabs are stored in 1 ml of sterile saline solution or in UTM Kit universal transport medium (Copan); nasopharyngeal aspirates and bronchoalveolar lavage are collected in sterile tubes for up to 72 h before processing (*see* **Note 2**).

3. The clinical samples should be transported at R.T. as fast as possible or refrigerated (2–8 °C) if longer time is required.

4. Pretreatment with Sputasol (Oxoid) is recommended in case of viscose samples (*see* **Note 3**).

2.2   DNA Extraction

1. Microcentrifuge.

2. Disposable plastic tubes (0.2 and 1.5 ml).

3. Disposable plastic filter tips (1–30 µl; 1–200 µl; 1–1.000 µl Eppendorf tips).

4. P20, P200, and P1000 micropipettes (*see* **Note 4**).

5. Sterile saline solution.

6. DNA extraction reagents (*see* **Note 5**).

7. Nucleic acid extraction robot (*see* **Note 6**).

8. Gloves (*see* **Note 7**).

9. Heating block.

2.3   qPCR Analysis

1. Scorpion probe-modified forward and classical reverse primers (Table 1) (*see* **Notes 8** and **9**).

2. Ready to use Master Mix (Invitrogen).

3. Tips and micropipettes for PCR (*see* **Note 10**).

4. Plastic tubes for reagent storage and master mix preparation (1.5 ml Eppendorf tubes).

Table 1
Real-time PCR primers and molecular Scorpion probes for the detection of *Mycoplasma pneumoniae*

ScoMycpn forward primer/scorpion probe
5′(FAM)-CggCggggTgCgTACAATACCATCCgCCg-(BHQ1)-(blocker)-gCCgCAAAgATgAAYgACg
Mycpn P1 reverse primer
5′-TCCTTCCCCATCTAACAgTTCAg-3′
β-actin forward primer
5′-ggggCTgTgCTgTggAAg-3′
β-actin reverse primer/scorpion probe
5′(Joe)-CgCgCTgCATTgCCgACAggATgAgCgCg(BHQ1)(Spacer—C18) gCCAgggCAgTgATCTCC-3′

The underlined bases correspond to the "probe" sequences complementary to the specific gene of interest: Mycoplasma P1 cytoadhesin type 1 and 2 and β-actin gene used as extraction control

5. 200 µl optical flat tubes.

6. Gloves.

7. Thermal cycler (Rotor Gene 3000 Qiagen).

8. Mycoplasma pneumoniae stain DNA (NCTC 010119), Minerva Biolabs.

3   Methods

Carry out all procedures at room temperature (R.T.) unless otherwise specified.

3.1   Sample Extraction

1. The swabs collected in 1 ml sterile saline solution or in 1 ml UTM Kit universal transport medium (Copan) are centrifuged at $16,800 \times g$ for 10 min. Pellets are resuspended in 200 µl of PBS and processed to extract DNA by QIAamp DNA Mini Kit.

2. In order to avoid cross-contamination automatic extraction by bio-robots is recommended when many samples have to be processed together. In this case it is possible to start directly from 1 ml of swab using NucliSENS easyMAG Biomérieux or 200 µl of final volume choosing Magtration System 12GC Plus and Magtration reagent (MagDEA Viral DNA/RNA 200 GC). Nasopharyngeal aspirates and bronchoalveolar lavage specimens are processed as nasopharyngeal swabs without any liquid addition.

3. In case of sample viscosity treat it with Sputasol as indicated (*see* **Note 3**).

4. Before extraction 1 µg of carrier DNA (from herring sperm DNA) is added to the sample to improve the DNA precipitation when a low amount of cell is present in it.

5. All the samples are eluted in 50 µl and stored at −20 °C.

3.2 Real-Time PCR Primers and Probes

1. Primers and unimolecular probes targeting either the P1 cytoadhesin type 1 and 2 gene of the Mp genome (i.e., AF286371, AF290001, and homologous sequences) or the β-actin sequence, used as extraction positive control, are shown in Table 1 (*see* **Notes 8** and **9**).

3.3 Real-Time PCR Assay Conditions

The reaction was performed in 200 µl plastic flat-cap tubes. A final volume of 25 µl contains template DNA, Platinum Quantitative PCR SuperMix-UDG Master Mix (Invitrogen, Milano, Italy) (*see* **Note 11**), and reaction buffer 1× (200 µM dATP, dCTP, dGTP, 400 µM dUTP, 3 mM magnesium chloride, 0.75 U Platinum Taq DNA polymerase, 0.5 U UDG, 100 mM KCl, 40 mM Tris–HCl pH 8.4, and stabilizers). Amplification was performed using both ScoMycpn Forward primer/scorpion probe and Mycpn P1 Reverse Primer at 300 nM in the presence of the β-actin forward primer and the reverse primer/scorpion probe used at 50 and 100 nM, respectively (*see* **Note 12**). The amplification has been performed on Rotor Gene 3000 instruments (Corbett Research, Diatech SRL, Italy) with the thermodynamic profile of 40 cycles of denaturation at 95 °C for 10 s and annealing/extension step at 55 °C for 35 s (*see* **Note 13**). The normalized fluorescent signal (ΔRn) is automatically calculated by a computer algorithm that normalizes the reporter emission signal (*see* **Note 14**).

4   Notes

1. All the samples are collected upon hospital admission and before any antibiotic therapy administration.

2. Store any leftover specimens at −20 °C. Avoid repetitive freezing and thawing of the sample, because it may lead to degradation of nucleic acid and to decrease of the sensitivity of the assay.

3. One vial of commercially available Sputasol solution (7.5 ml) contains dithiothreitol 0.1 g, sodium chloride 0.78 g, potassium chloride 0.02 g, disodium hydrogen phosphate 0.112 g, and potassium dihydrogen phosphate 0.02 g dissolved in 7.5 ml distilled water (pH 7.4 ± 0.2 at 25 °C). The 7.5 ml vial is diluted in 92.5 ml of sterile distilled water, stored at 2–8 °C, and used within 48 h in a ratio volume 1:1 with the sample. Pipette up and down to homogenize samples. If needed perform a quick incubation at 37 °C to reduce viscosity.

4. It is important to perform sample check-in and sample extraction in a confined area where PCR reactions are not assembled. Use separated and segregated working areas for each process. Workflow in the laboratory should proceed in a unidirectional manner.

5. The nucleic acid is extracted using QIAamp DNA Mini Kit (QIAGEN spa, Milano, ITALY) following the manufacturer's instructions for blood and body fluid samples.

6. High-throughput automatic nucleic acid extraction is performed by NucliSENS easyMAG Biomérieux or Magtration System 12GC plus with Magtration reagent (MagDEA Viral DNA/RNA 200 GC).

7. Always wear disposable powder-free gloves in each area, and change them quite often during each working process to decrease the possibility of personal/sample contaminations.

8. Real-time PCR assay was developed by targeting the *Mycoplasma pneumoniae* P1 cytoadhesin type 1 and 2 sequences (AF286371, AF290001), searching among GenBank-available sequences for conserved region. In addition, to confirm the extraction of a valid biological template in each tube sample, we included primer and probe mix to detect the endogenous β-actin gene. Primers and unimolecular Scorpion probes were designed using both Primer Express (PE Biosystem, Foster City, CA) and Oligo 4.1 primer analysis software (National Biosciences Inc., Plymouth, MN) to select the best thermodynamically performing sequences.

9. Unimolecular Scorpion probe is essentially a bi-labeled fluorescent probe/primer hybrids with a nucleotide sequence region complementary to the same target gene [8]. It carries the probe element at the 5′ end of the nucleotide, and in standard condition it is thermodynamically stable ("off-conformation"), since it displays a hairpin loop conformation by the presence of a self-complementary 6 bp stem sequence at the 5′ and 3′ ends. In this conformation the reporter and quencher are close enough (<20 nM) that, following the Förster resonance energy transfer (FRET) principle, the fluorescence emission is efficiently quenched [9]. After each cycle of PCR a new DNA complementary target region will be synthesized, and following the successive steps of denaturation and annealing, the probe will hybridize to a part of the newly produced PCR product. The opening of the unimolecular probe loop and its hybridization have a thermodynamically favorable kinetics (differential free energy of at least −2.0 kcal/mol) that leads to the separation of the fluorophore from the quencher and causes light emission. The thermodynamic modeling needed to design such a probe has been possible

using the DNA *mfold* suite on the Michael Zuker Web site [10] according to the thermodynamic parameters established by John Santalucia [11].

10. Dedicate both filter tips and micropipettes for the separate working areas (i.e., nucleic acid isolation, reagent mixing, and nucleic acid template addition) in order to prevent splashing and cross-contamination. Do not move them from one area to another.

11. Carryover contamination between PCR reactions can be prevented by including uracil-*N*-glycosylase (UNG) in each tube supplemented with reagents. Thus, some commercial PCR pre-made mixes may already contain UNG, or alternatively it can be added as a separate component. UNG can only prevent carryover from PCR reactions (PCR-derived cross-contamination), since the amplification products include deoxyuridine triphosphate (dUTP) that may be degraded before starting with the PCR reaction (15-min preincubation step at 37 °C using 0.2 U/tube of enzyme); UNG is then inactivated at 95 °C during the first PCR step.

12. Upon arrival primers and probes are resuspended in sterile water at the concentration of 25 μM. The working concentration is 5 μM for both of them. They are aliquoted in small volume and stored at −20 °C.

13. Any real-time PCR run requires a positive and a negative control. We use *Mycoplasma pneumoniae* (NCTC 010119), Minerva Biolabs, as positive control and sterile water as negative control; the control of extraction is traced in each tube by the β-actin assay that hybridizes to the human genomic DNA always contaminating each extracted sample.

14. Sample positivity is evaluated when it reaches, upon PCR amplification, the fluorescence emission of the fixed threshold value that is maintained identical in all the sets of experiment of an array.

References

1. Blasi F, Tarsia P, Aliberti S et al (2005) *Chlamydia pneumoniae* and *Mycoplasma pneumoniae*. Semin Respir Crit Care Med 26: 617–624

2. Talkington DF, Shott S, Fallon MT et al (2004) Analysis of eight commercial enzyme immunoassay tests for detection of antibodies to *Mycoplasma pneumoniae* in human serum. Clin Diagn Lab Immunol 11:862–867

3. Mackay IM (2004) Real-time PCR in the microbiology laboratory. Clin Microbiol Infect 10:190–212

4. Welti M, Jaton K, Altwegg M et al (2003) Development of a multiplex real-time quantitative PCR assay to detect *Chlamydia pneumoniae*, *Legionella pneumophila* and *Mycoplasma pneumoniae* in respiratory tract secretions. Diagn Microbiol Infect Dis 45:85–95

5. Stralin K, Backman A, Holmberg H et al (2005) Design of a multiplex PCR for *Streptococcus pneumoniae*, *Haemophilus influenzae*, *Mycoplasma pneumoniae* and *Chlamydophila pneumoniae* to be used on sputum samples. APMIS 113: 99–111

6. Wong ML, Medrano JF (2005) Real-time PCR for mRNA quantitation. Biotechniques 39:75–85

7. Whitcombe D, Theaker J, Guy SP et al (1999) Detection of PCR products using self-probing amplicons and fluorescence. Nat Biotechnol 17:804–807

8. Di Marco E, Cangemi G, Filippetti M et al (2007) Development and clinical validation of a real-time PCR using uni-molecular Scorpion-based probe for the detection of Mycoplasma pneumoniae in clinical isolates. New Microbiol 30:415–421

9. Jie Z (2006) Spectroscopy-based quantitative fluorescence resonance energy transfer analysis. In: Stockand JD, Shapiro MS (eds) Ion channels: methods and protocols, methods in molecular biology, vol 337. Humana, Totowa, NJ, pp 65–77

10. Zuker M (2003) Mfold web server for nucleic acid folding and hybridization prediction. Nucleic Acids Res 31:3406–3415

11. SantaLucia J Jr (1998) A unified view of polymer, dumbbell, and oligonucleotide DNA nearest-neighbor thermodynamics. Proc Natl Acad Sci U S A 95:1460–1465

Chapter 10

A Sensible Technique to Detect Mollicutes Impurities in Human Cells Cultured in GMP Condition

Elisabetta Ugolotti and Irene Vanni

Abstract

In therapeutic trials the use of manipulated cell cultures for clinical applications is often required.

Mollicutes microorganism contamination of tissue cultures is a major problem because it can determine various and severe alterations in cellular function.

Thus methods able to detect and trace cell cultures with Mollicutes contamination are needed in the monitoring of cells grown under good manufacturing practice conditions, and cell lines in continuous culture must be tested at regular intervals.

We here describe a multiplex quantitative polymerase chain reaction assay able to detect contaminant Mollicutes species in a single-tube reaction through analysis of 16S–23S rRNA intergenic spacer regions and Tuf and P1 cytoadhesin genes.

The method shows a sensitivity, specificity, and robustness comparable with the culture and the indicator cell culture as required by the European Pharmacopoeia guidelines and was validated following International Conference on Harmonization guidelines and Food and Drug Administration requirements.

Key words Multiplex qPCR, European Pharmacopoeia, Good manufacturing practice, Mollicutes, Mycoplasmas, Acholeplasmas, Tissue culture contaminants

1 Introduction

In therapeutic trials the use of manipulated cell cultures and their precursors for clinical applications is often required.

Patients with malignancies and hematopoietic disorders or undergoing CMV or EBV infections may benefit from the treatment with manipulated and/or expanded virus-specific T lymphoid cells [1] that must be constantly subjected to microbiological monitoring.

Indeed reinfused material needs careful microbial surveillance and to be grown in a good manufacturing practice (GMP) environment, following the European Directive, Food and Drug Administration (FDA) requirements, and International Conference on Harmonization (ICH) guidelines [2–5].

Roberto Biassoni and Alessandro Raso (eds.), *Quantitative Real-Time PCR: Methods and Protocols*, Methods in Molecular Biology, vol. 1160, DOI 10.1007/978-1-4939-0733-5_10, © Springer Science+Business Media New York 2014

Contamination of tissue culture is frequently observed and is awkward to prevent because it may be operator induced or linked to cell culture medium recipes.

The contaminants most frequently found in cell culture are the Mollicutes that being small and without cell wall are difficult to eradicate and to detect with conventional microbiological methods [6–10].

Mollicutes represents a large group of highly specialized bacteria, but only a limited number of Mycoplasma as *M. fermentans, M. pneumoniae, M. orale, M. arginini* and *M. hyorhinis* [11, 12] and Acholeoplasma as *A. laidlawii* species occur predominantly in cell culture and are the most challenging to highlight [13].

To detect mycoplasma tissue culture contamination a wide spectrum of approaches have been proposed like molecular assay, enzyme immunoassay, microbiological culture, and direct/indirect DNA staining [14–16], but nucleic acid amplification techniques (NAT) represent an efficient alternative detection system.

Indeed PCR assay when validated according to European Pharmacopoeia (EuPh) guidelines 2.6.7 Mycoplasmas [17] is able to reach a sensitivity, specificity, robustness, and simplicity comparable with either the cell culture or the indicator cell culture method [16].

Moreover the NAT application in biologic products may improve the efficiency of detection allowing the identification of the different mycoplasma types with relatively low time and labor effort combined with high analytical sensitivity.

Among NAT, the quantitative polymerase chain reaction (qPCR) once optimized is the best method as it provides the highest levels of sensitivity without the need for confirmatory tests [12].

Essential conditions required for the NAT validation are the following: (1) the detection limit ≤ 10 colony-forming units (CFU)/ml; (2) the species tested must be *A. laidlawii, M. fermentans, M. pneumoniae, M. orale, M. arginini*, and *M. hyorhinis*; and (3) specificity of detection that is reached by exclusion of the phylogenetically close bacteria such as Lactobacillus, Clostridium, and Streptococcus.

Here we present a method using multiplex qPCR detection system. It is able to identify Mollicutes species that may contaminate cell cultures under GMP conditions, and it may be useful for clinical applications. Using primers specific for the 16S–23S rRNA intergenic spacer regions, for tuf gene, and for P1 cytoadhesin [18–24] the method is able to detect the most common tissue culture contaminant species in a single-tube reaction complying the sensitivity, specificity, and robustness required by EuPh guideline.

2 Materials

2.1 Mollicutes DNA Extraction from Tissue Culture Supernatants

1. NucliSENS easyMAG lysis buffer (bioMerieux, Durham, NC).
2. NucliSENS easyMAG extraction buffer 1, 2, 3 (bioMerieux, Durham, NC).
3. NucliSENS easyMAG magnetic silica (bioMerieux, Durham, NC).
4. NucliSENS easyMAG disposables (bioMerieux, Durham, NC).
5. NucliSENS easyMAG instrument (bioMerieux, Durham, NC).

2.2 Microorganism Genomic DNA and Internal Control DNA

1. *M. fermentans, M. pneumoniae, A. laidlawii* certificated titled DNA standards (1×10^6 genomes/μl) (Minerva Biolabs GmbH, Berlin, Germany) were diluted and used as positive control (*see* **Note 1**).
2. Synthetic 69-mer DNA fragment of beta globin 5′-TGA GCC AGG CCA TCA CTA AAG GCA CCG AGC ACT TTC TTG CCA TGA GCC TAG AAC CTC TGG GTC CAA GGG-3′ (TIB BioMol s.r.l., Italy) was used as internal control in the working solution of 100 copies/μl [18].

2.3 Multiplex qPCR

1. Primers and probes (*see* **Notes 2** and **3**):
   The primer and probe sequences and their working concentration are shown in Table 1.
      Probes should be synthesized as described in Table 1.
2. EXPRESS qPCR Supermix Universal (Invitrogen).
3. ABI 7500 Instrument (Applied BioSystems).
4. MicroAmp™ Optical 96-Well Reaction Plate (Applied BioSystems) (*see* **Note 4**).
5. MicroAmp™ Optical Adhesive Film (Applied BioSystems).
6. Nuclease-free water.
7. Microcentrifuge and vortex for mixing preps.
8. Tubes RNase, DNase, DNA, and PCR inhibitor free.
9. MicroPipettes (single- and multichannel).
10. Filter tips (RNase, DNase, DNA, and PCR inhibitor free).
11. Disposable gloves.
12. Biocontainment hoods (*see* **Note 5**).

3 Methods

3.1 DNA Sample Extraction (See Note 6)

1 μl of the synthetic oligo-deoxynucleotide (beta globin) solution (100 copies/μl) has been added in each sample before DNA extraction procedure (*see* **Note 7**).

Table 1
Mollicutes real-time PCR primers and probes

Primers and probes	Sequence 5'–3'	Working concentration (μM)	Final concentration (nM)	Gene target	Species
b-glob frw	TGA GCC AGG CCA TCA CTA AAG	60	300	Beta globin	*Homo sapiens*
b-glob rev	CCC TTG GAC CCA GAG GTT CT	60	300		
b-glob TQ probe	Cy5-CAC CGA GCA CTT TCT YGC CAT GAG C-BBQ	40	200		
Al frw	ATT ACG TGC TAC TGA CAA ACC ATT TA	50	250	Elongation factor gene (TUF)	*A. laidlawii*
Al rev	GAT CAA CAC GTC CTG TAG CAA CT	50	250		
AlP1MGB probe	FAM-CAC GAC CTG TAA TTG TG-NFQ	40	200		
MF2 frw	AAT YTG CCG GGA CCA CC	60	300	16S–23S rRNA intergenic spacer regions	FOAHS
MR1 rev	CCT TTC CCT CAC GGT ACT AG	60	300		
MoP2 LNA probe	FAM-TT+C A+CT AT+C GGT GT+C TG-BBQ	40	200		
Mycpn P1-F	GCC GCA AAG ATG AAY GAC G	60	300	P1 cytoadhesin gene	*M. pneumoniae*
Mycpn P1-R	TCC TTC CCC ATC TAA CAG TTC AG	60	300		
Dual-labeled probe	FAM-TTG ATG GTA TTG TAC GCA CCC CAC TCG-BBQ	40	200		

FOAHS: *M. fermentans, M. orale, M. arginini, M. hyorhinis, M. salivarium*
+C means LNA-modified C nucleotides
Y can be C/T (IUPAC code)
References [23, 24]
A. laidlawii detection was performed using a minor groove binder (MGB) probe labeled by 5' FAM and a 3' black hole quencher 1 (BHQ1)
M. pneumoniae detection was performed using a dual-labeled probe by 5' FAM and a 3' BHQ1
FOAHS detection was performed using a specific lock nucleic acid (LNA) probe labeled by 5' FAM and a 3' blackberry quencher (BBQ)
Beta globin internal control detection was performed using a dual-labeled probe by 5' CY5 and a 3' BBQ

Three (1 ml each) aliquots of culture media supernatant for each sample must be extracted using NucliSENS easyMAG instrument based on a magnetic silica particle purification protocol (*see* **Note 8**). DNA extraction is carried out according to the manufacturer's protocol [25], eluting the DNA template in 40 µl (for each 1 ml of original sample) of elution buffer.

The whole eluted single sample is used for each qPCR determination.

3.2 Multiplex qPCR Procedure

The used PCR cycling was as follows:

Pretreatment for 2 min at 50 °C with uracil-DNA glycosylase (UDG) followed by 2 min at 95 °C followed by 40 cycles at 95 °C for 15 s and at 60 °C for 35 s (*see* **Note 9**).

For each DNA sample, the multiplex reaction was assembled using 60 µl of master mix (*see* **Note 10**), the total volume of the eluted sample, and water to obtain a final volume of 100 µl.

The 60 µl of master mix was produced by adding 50 µl of Express qPCR Supermix Universal to 0.5 µl of working concentration for each primer and probe as described in Table 1 (8 primers + 4 probes for a volume of 6 µl) and 3.8 µl of nuclease-free H_2O and 0.2 µl of ROX (*see* **Note 11**).

Each RQ-PCR plate must include no-template controls (40 µl of nuclease-free H_2O, add to master mix instead of DNA sample); negative controls (40 µl of no-Mollicutes tissue culture supernatant, subject to the same treatment of clinical sample, add to master mix); positive controls (40 µl containing *M. fermentans*, *M. pneumoniae*, *A. laidlawii* DNA standards for a total copies number of 1×10^3 obtained adding 10 µl of three diluted Mollicutes DNA standard and 10 µl of nuclease free H_2O); and clinical specimens.

All control and samples should be tested in triplicates.

Load the samples in the plate in the order previously established in the work plan of the software Applied Biosystems Sequence Detection Software (SDS).

After the loading samples in the 96-well reaction plate, seal the reaction plate with an optical adhesive film, put the plate onto the instrument, and run the assay.

When the run has completed select "Analyze" to set a threshold value, save the results, and remove the reaction plate from the instrument.

3.3 Result Analysis

First you must set the threshold value so that the line crosses the curves of positive controls at the beginning of the exponential phase (*see* **Note 12**).

All controls should be analyzed first to validate the experiment.

Check that the no-template controls and negative controls resulted negative (*see* **Note 13**).

The positive controls should intercept the threshold around the 30th cycle (*see* **Note 14**) and beta globin internal control should be positive (*see* **Note 15**).

The clinical samples are considered positives when an exponential curve crosses the threshold value (*see* **Note 16**).

Finally specimen information, threshold data, and Ct value obtained in the test may be displayed in a report file generated using the software instrument.

4 Notes

1. Each DNA standard is diluted $1:3 \times 10^4$ in nuclease-free water in order to obtain 33.3 copies/µl.

2. Primer and probe purity is crucial.

    Primers should be manufactured with standard quality and probes by HPLC purification, and the working solutions should be prepared by diluting in nuclease-free molecular-grade water.

    Working solutions should be maintained at 4 °C; batches should be aliquoted into small volumes for one-time use and frozen at −20 °C. Repeated freeze–thaw cycles are not recommended for stability purposes.

    Probes should be protected from light to avoid degradation of the probe fluorophore.

3. In order to increase the detection specificity a combination of LNA, MGB, and dual-labeled probes should be used. LNA probe uses a nucleic acid analogue (containing a 2′-O, 4′-C methylene bridge in the pentose structure) that increases thermal stability and hybridization specificity and enables the design of shorter sequences than standard probes [26]. MGB probe should be used because it is able to guarantee a high specificity in the presence of a single mismatch [18–20].

4. The covers primarily require the application of pressure by the user to ensure a tight, evaporation-free seal. Improper peeling of the cover may result in haziness but does not affect results.

5. To prevent contamination of samples during the preparation of the 96 wells plate and the handling of the different mollicutes DNA.

6. For each biological sample it is recommended to have available at least three aliquots of 1 ml of culture media supernatant which will be used to perform the test in triplicate.

    The crude cell culture supernatants can be stored at 4 °C for a few days or frozen at −20 °C for several weeks. After thawing, the samples should be further processed immediately.

7. For routine verification of the absence of inhibition and for evaluation of the loss of material during the DNA extraction steps always insert an internal control in the test.

8. Here we present an automated extraction method to process different samples, but you may use any manual DNA extraction

procedure maintaining the final volume of elution as stated for automatic procedure.

9. If the used thermal cycler needs the use of ROX as passive reference dye remember to switch off its detection.

10. It is important that separate and dedicated laboratory rooms must be used for different stages of processing. So, it is necessary to maintain a separate "clean" room for all reagents and consumables from a "dirty" room used for DNA addition. These practices prevent contamination of laboratory.

11. The final volume of master mix is based on the total number of reactions required for each plate plus two additional volumes.

12. The default displays the data in a logarithmic format, but it may be more easily visualized on a linear scale.

13. These controls should not possess an exponential growth curve within 40 cycles; otherwise, they are indicative of Mollicutes contamination and so the assay is invalid.

14. Lack of amplification curve on any samples including the positive controls may indicate a problem in the preparation of the master mix.

15. The internal control value is helpful in assessing the extracted nucleic acid quality.

    A nonexistent or low amplification curve indicates poor-quality template, and the samples must be re-extracted.

    Particular attention must be paid to good laboratory practices because beta globin, being a marker that detects human DNA, easily contaminates material and equipment used for this analysis.

16. It is fundamental to examine the curve shape in particular for those who have late Ct values. Specimens with a Ct value >38 need to be examined carefully as they are suspected of being true-positive sample. However they may be an artefact. It is critical to also examine the "Spectra" and "Component" tabs to help with the analysis.

References

1. Li Pira G, Ivaldi F, Tripodi G et al (2008) Positive selection and expansion of cytomegalovirus-specific CD4 and CD8 T cells in sealed systems: potential applications for adoptive cellular immunoreconstitution. J Immunother 31:762–770

2. World Health Organization (1992) Good manufacturing practices for pharmaceutical products. In: WHO Expert Committee on Specifications for Pharmaceutical Preparations. 32nd report. WHO technical report series No. 823. World Health Organization, Geneva; Annex 1

3. European Union (2004) Directive 2004/23/EC of the European Parliament and of the Council on Setting Standards of Quality and Safety for the Donation, Procurement, Testing, Processing, Preservation, Storage and Distribution of Human Tissues and Cells. Official Journal of the European Union; Strasbourg, 31 March

4. Zoon KC (2005) Food and drug administration points to consider in the characterization of cell lines used to produce biologicals. 21. CFR 610.30. Food and Drug Administration, Bethesda, MD, USA. 1993. http://www.fda.

gov/cber/gdlns/ptccell.pdf. Accessed 24 May 2005

5. International Conference on Harmonization (1995) ICH Q2 (R1) validation of analytical procedure: text and methodology. In: Proceedings of the international conference on harmonization, Geneva. http://www.ich.org/fileadmin/Public_Web_Site/ICH_Products/Guidelines/Quality/Q2_R1/Step4/Q2_R1_Guideline.pdf

6. Razin S, Hayflick L (2010) Highlights of mycoplasma research: an historical perspective. Biologicals 38:183–190

7. Shahhosseiny MH, Hosseiny Z, Khoramkhorshid HR et al (2010) Rapid and sensitive detection of Mollicutes in cell culture by polymerase chain reaction. J Basic Microbiol 50:171–178

8. Uphoff CC, Drexler HG (2005) Detection of mycoplasma contaminations. Methods Mol Biol 290:13–23

9. Uphoff CC, Drexler HG (2005) Eradication of mycoplasma contaminations. Methods Mol Biol 290:25–34

10. Drexler HG, Uphoff CC (2002) Mycoplasma contamination of cell cultures: incidence, sources, effects, detection, elimination, prevention. Cytotechnology 39:75–90

11. Rawadi G, Dussurget O (1995) Advances in PCR-based detection of mycoplasmas contaminating cell cultures. PCR Methods Appl 4:199–208

12. Young L, Sung J, Stacey G et al (2010) Detection of Mycoplasma in cell cultures. Nat Protoc 5:929–934

13. Razin S, Yogev D, Naot Y (1998) Molecular biology and pathogenicity of mycoplasmas. Microbiol Mol Biol Rev 62:1094–1156

14. Loens K, Ursi D, Goossens H et al (2003) Molecular diagnosis of Mycoplasma pneumoniae respiratory tract infections. J Clin Microbiol 41:4915–4923

15. Harris R, Marmion BP, Varkanis G et al (1988) Laboratory diagnosis of Mycoplasma pneumoniae infection. II. Comparison of methods for the direct detection of specific antigen or nucleic acid sequences in respiratory exudates. Epidemiol Infect 101:685–694

16. Volokhov DV, Graham LJ, Brorson KA et al (2011) Mycoplasma testing of cell substrates and biologics: review of alternative non-microbiological techniques. Mol Cell Probes 25:69–77

17. European Directorate for the Quality of Medicines (EDQM), Council of Europe, Strasbourg, France. European pharmacopoeia 5.0. 2004 Section 2.6.7. Mycoplasma

18. Harasawa R (1999) Genetic relationships among mycoplasmas based on the 16S–23S rRNA spacer sequence. Microbiol Immunol 43:127–132

19. Harasawa R, Kanamoto Y (1999) Differentiation of two biovars of ureaplasma urealyticum based on the 16S–23S rRNA intergenic spacer region. J Clin Microbiol 37:4135–4138

20. McGarrity GJ, Kotani H (1986) Detection of cell culture mycoplasmas by a genetic probe. Exp Cell Res 163:273–278

21. Kong F, James G, Gordon S et al (2001) Species-specific PCR for identification of common contaminant Mollicutes in cell culture. Appl Environ Microbiol 67:3195–3200

22. Stormer M, Vollmer T, Henrich B et al (2009) Broad-range real-time PCR assay for the rapid identification of cell-line contaminants and clinically important Mollicutes species. Int J Med Microbiol 299:291–300

23. Di Marco E, Cangemi G, Filippetti M et al (2007) Development and clinical validation of a real-time PCR using uni-molecular Scorpion-based probe for the detection of Mycoplasma pneumoniae in clinical isolates. New Microbiol 30:415–421

24. Vanni I, Ugolotti E, Raso A et al (2012) Development and validation of a multiplex quantitative polymerase chain reaction assay for the detection of Mollicutes impurities in human cells, cultured under good manufacturing practice conditions, and following European Pharmacopoeia requirements and the international conference on harmonization guidelines. Cytotherapy 14:752–766

25. Boom R, Sol CJ, Salimans MM et al (1990) Rapid and simple method for purification of nucleic acids. J Clin Microbiol 28:495–503

26. Kaur H, Arora A, Wengel J et al (2006) Thermodynamic, counterion, and hydration effects for the incorporation of locked nucleic acid nucleotides into DNA duplexes. Biochemistry 45:7347–7355

Chapter 11

Real-time Quantification Assay to Monitor *BCR-ABL1* Transcripts in Chronic Myeloid Leukemia

Pierre Foskett, Gareth Gerrard, and Letizia Foroni

Abstract

The *BCR-ABL1* fusion gene, the causative lesion of chronic myeloid leukemia (CML) in >95 % of newly presenting patients, offers both a therapeutic and diagnostic target. Reverse-transcription quantitative polymerase chain reaction technology (RT-qPCR), utilizing primer–probe combinations directed to exons flanking the breakpoint junctional region, offers very high levels of both specificity and sensitivity, in a scalable, robust, and cost-effective assay.

Key words BCR-ABL1, RT-qPCR, Real-time PCR, Quantification, Reverse transcription, cDNA, CML

1 Introduction

The molecular hallmark of chronic myeloid leukemia (CML) and Philadelphia positive acute lymphoblastic leukemia (Ph+ALL) is the *BCR-ABL1* fusion gene. This is the consequence of a t(9;22) (q34;q11) translocation event, which within the context of a single hematopoietic stem cell gives rise to the bulk disease through clonal expansion [1]. The resultant BCR-ABL1 oncoprotein forms a homodimer that through autophosphorylation acts as a potent dysregulated tyrosine kinase. This signal, through multiple pathways, affects cellular proliferation, adhesion, and apoptosis [2], particularly of the myeloid cells, but often affects all lineages.

Regular and accurate monitoring is of particular importance in CML. Since molecular milestones have become ever more important for informing clinical management decisions it is paramount that accurate molecular monitoring is achieved, especially in the context of switching between tyrosine kinase inhibitor/s (TKIs) in response to suboptimal efficacy at early time points [3] or loss of response because of resistance or poor adherence [4].

There are several *BCR-ABL1* isoforms that differ in the location of the breakpoint junction regions between the two genes

Roberto Biassoni and Alessandro Raso (eds.), *Quantitative Real-Time PCR: Methods and Protocols*, Methods in Molecular Biology, vol. 1160, DOI 10.1007/978-1-4939-0733-5_11, © Springer Science+Business Media New York 2014

(conventionally referred with "e" for BCR and "a" for ABL1) and between the fusion exons and follow the nomenclature e$Xa\Upsilon$, where X and Υ are the BCR and ABL1 exons proximal to the breakpoint junction, respectively. The breakpoint regions are mostly intronic and for ABL1 on chromosome 9 they almost always lead to the formation of a transcript with exon 2 (a2) being 3' and proximal to the breakpoint junction. Rare variants exist where the breakpoint is downstream of exon 2 and exon 3 is proximal (a3). Within the BCR gene on chromosome 22, there are two common breakpoint cluster regions: the "major" breakpoint region (associated with CML), which lies between exons 12 and 16, giving rise to the e13a2 and e14a2 isoforms (p210) and the "minor" region (associated with Ph positive ALL), which lies between exons 1 and 2 of the BCR gene, leading to the e1a2 (p190) fusion [5]. Other rare breakpoint regions form species, which result in the BCR exon 6, 8, and 19 being proximal to the junctional region (e6a2, e8a2, and e19a2, respectively). Hypothetically, all of these isoforms could also exist as an a3 variant, but in practice, only the e13a3, e14a3 (and very occasionally the e1a3) have been so far described [6].

It is of paramount importance that a patient's disease associated breakpoint is correctly characterized at diagnosis (typically by multiplex endpoint PCR [7]), before molecular monitoring by reverse transcription quantitative PCR (RT-qPCR) can be applied to follow up samples. The short-amplicon nature of RT-qPCR means that prime-probe sets used for each breakpoint species differ in at least one primer, which incorrectly applied could lead to false-negative results. Since the majority of CML samples seen by individual labs will be almost exclusively restricted to the classical e13/e14a2 variety we have described our protocol for quantification of this transcript type, and it may be advisable that monitoring patients with rare transcripts is done within a specialist center, like ours.

The primary references for the RT-qPCR workflow for the molecular monitoring of BCR-ABL1 associated malignancies used by the majority of involved centers, at least in Europe, are those produced by the Europe Against Cancer initiative [8, 9]. These were later updated and compiled into a UK guidelines manuscript [10] and further refinements were made (including automation, duplex probes, and fast-mode cycling) to optimize for scalability and high throughput [11]. The protocol described herein details how to extract RNA from whole peripheral blood, obtain the necessary high quality cDNA through reverse transcription and perform a duplex RT-qPCR for the BCR-ABL1 e13/14a2 transcripts (indiscriminate between e13a2 and e14a2) and ABL1 control gene to produce a ratio which represents disease burden. This is currently the default workflow for our center (Imperial Molecular Pathology, Hammersmith Hospital, London, UK).

2 Materials

2.1 Total White Blood Cell Isolation and Lysis

1. Red Cell Lysis buffer, to make 5 l weigh 41.5 g of ammonium chloride (NH_4Cl) and 5 g of potassium bicarbonate ($KHCO_3$). Add to 4 l of distilled water. Add 1 ml of 0.5 mol/l EDTA. Allow solids to dissolve. Make up to 5 l with distilled water. Adjust pH to 7.4 using HCl. Store at 4 °C.

2. Phosphate buffered saline (PBS).

3. RNeasy Mini Kit (Qiagen)—contains RLT.

4. 2-mercaptoethanol.

5. 50 ml capped polypropylene tubes.

6. Centrifuge (capable of spinning the 50 ml tubes at $400 \times g$).

7. 2 ml Sample Tubes.

8. 2 ml Syringes.

9. Blunt ended 18G 1½″ needle.

2.2 RNA Extraction

1. RNeasy Mini Kit (Qiagen).

2. 70 % ethanol.

3. 1.5 ml microcentrifuge capped tubes.

2.3 cDNA Synthesis

Each 100 μl of reverse transcriptase reaction is made up of:

1. 55 μl of the eluted RNA.

2. 45 μl of cDNA reagent mix:
   - 20 μl 5× Buffer.
   - 10 μl (0.1 M) DTT.
   - 2 μl 25 mM dNTPs.
   - 0.2 μl 3 μg/μl random hexamer primers.
   - 9.2 μl ddH$_2$O.
   - 2.4 μl MMLV enzyme (200 U/μl).
   - 1.2 μl of RNasin (20 U/μl).

2.4 RT-qPCR

1. qPCR thermal cycler with fast mode (i.e., ABI—Life Technologies ViiA, StepOnePlus, 7900HT, or 7500FAST. Other qPCR platforms may be used, but should be subject to validation of optimal conditions).

2. MicroAmp Fast Optical 96-well reaction plate (Life Technologies).

3. MicroAmp Optical plate seals (Life Technologies).

4. TaqMan Fast Advanced Master Mix (Life Technologies).

5. Primer–probe mix:

Table 1
Primer–probe mix calculations

	Per well (add 7 µl)	For 1 × 96-well plate (make for 110 runs to allow for pipetting errors)
Primers		
ENF501 (80 µM)	0.075 µl	8.3 µl
ENF561 (80 µM)	0.075 µl	8.3 µl
ENF1003 (80 µM)	0.038 µl	4.1 µl
ABL1063 (80 µM)	0.038 µl	4.1 µl
Probes		
ENP541F-MGB (100 µM)	0.02 µl	2.2 µl
ABL1043V-MGB (100 µM)	0.04 µl	4.4 µl
dH$_2$O	6.72 µl	738.7 µl

Primer sequences

- ENF501: TCCGCTGACCATCAAYAAGGA (Y=any pyrimidine).

- ENF561: CACTCAGACCCTGAGGCTCAA.

- ENF1003: TGGAGATAACACTCTAAGCATAACTAA AGGT.

- ABL1063: GATGTAGTTGCTTGGGACCCA.

Minor Groove Binding (MGB) probe sequences

- ENP541F-MGB: 6FAM-CCCTTCAGCGGCCAGT.

- ABL1043V-MGB: VIC-CATTTTTGGTTTGGGCTTC.

6. Each 20 µl of PCR reaction is made up of the following components:

   - 10 µl of 2× TaqMan Fast Advanced Master Mix (Life Technologies).

   - 3 µl of cDNA.

   - 7 µl of primer–probe mix (*see* Table 1):

7. Plasmid: Certified *BCR-ABL1* pDNA CALIBRANT (IRMM) available at 1×10^1, 1×10^2, 1×10^3, 1×10^4, 1×10^5, 1×10^6 copies per 1 µl.

3 Methods

3.1 Total White Blood Cell Isolation and Lysis

1. 10–20 ml whole blood is collected from the patient using EDTA as an anticoagulant (*see* **Note 1**).

2. Transfer 10–20 ml whole blood in EDTA into the 50 ml polypropylene tube (*see* **Note 2**) and add Red Cell Lysis buffer to fill the tube to 45 ml final volume.

3. Secure the cap (*see* **Note 3**) and leave on ice for 10 min.

4. Centrifuge tubes at $400 \times g$ for 7 min on a bench centrifuge then carefully pour off the supernatant retaining the white cell pellet at the bottom of the tube. Vortex to break the pellet.

5. Add Red Cell Lysis Buffer to 40 ml, secure the cap and leave on ice for an additional 10 min.

6. Centrifuge tubes at $400 \times g$ for 7 min then carefully pour off the supernatant retaining the white cell pellet at the bottom of the tube.

7. Add phosphate buffered saline (PBS) up to 30 ml, secure the cap.

8. Centrifuge tubes at $400 \times g$ for 7 min then carefully pour off the supernatant retaining the white cell pellet at the bottom of the tube.

9. Add 1 ml of RLT containing 10 μl of beta-mercaptoethanol.

10. Pipette with a plastic Pastette to break the pellet and transfer the lysate to a 2 ml sample tube.

11. Homogenize lysate by repeated passes through an 18G blunt needle by syringe until the solution loses its viscosity (*see* **Notes 4** and **5**).

12. Freeze at –20 °C overnight or at –80 °C indefinitely (*see* **Note 6**).

3.2 RNA Extraction

1. Transfer 350 μl of the RLT lysate to a 1.5 ml microcentrifuge tube and add 350 μl of 70 % ethanol (*see* **Note 7**).

2. Transfer the 700 μl mix to an RNeasy spin column arranged in a 2 ml collection tube then centrifuge for 15 s at $10,000 \times g$ and discards the flow-through.

3. Add 650 μl RW1 buffer to the spin column and centrifuge for 15 s at $10,000 \times g$.

4. Discard the flow-through and replace the 2 ml collection tube, then add 500 μl RPE wash buffer to the spin column and centrifuge for 15 s at $10,000 \times g$, discarding the flow-through afterwards.

5. Add 500 μl RPE washing buffer to the spin column and centrifuge for 2 min at $20,000 \times g$, then transfer the spin column to a 1.5 ml collection tube and allow the columns to air-dry for 20 min.

6. Add 60 μl of RNase free water to the spin column and elute the RNA by centrifuging for 2 min at $20,000 \times g$, the spin column can then be discarded (*see* **Note 8**).

3.3 cDNA Synthesis

1. Incubate 55 μl of each RNA eluate at 65 °C for 10 min, then immediately transfer the tubes to ice for 30 s (*see* **Note 9**).

2. Pulse spin tubes to draw contents to the bottom and to each sample add 45 μl of cDNA reagent mix, combining the solutions by gently pipetting (*see* **Notes 10** and **11**).

3. Incubate tubes at 37 °C for 2 h.

4. Stop the reaction by incubating tubes at 65 °C for 10 min followed by a pulse spin to draw contents to the bottom of the tubes (*see* **Note 12**).

3.4 Quantitative PCR (RT-qPCR)

1. Draw up a layout of a 96-well PCR plate accounting for analysis of each sample, and each plasmid, in triplicate (*see* **Note 13**). At least three dilutions of the plasmid should be used, preferably at concentrations of 100,000, 10,000, and 1,000 molecules.

2. To each well in use add:

   - 10 µl Taqman Fast Advanced Master Mix.

   - 7 µl of primer–probe mixture (*see* **Notes 14** and **15**).

3. Add 3 µl of cDNA to each designated (triplicate) well (*see* **Note 16**).

4. In a physically separated area (preferably a completely separate room), using different pipettes and tips, add 3 µl of plasmid to each designated triplicate well.

5. Carefully seal the top of the plate with optical plate seal.

6. Pulse-spin the plate to draw contents to the bottom.

7. Place plate in the real-time PCR machine using the program: 95 °C for 20 s followed by 45 cycles of: 95 °C for 3 s and 60 °C for 45 s. With settings of Fast Mode (modified), Ramp Rate: 100 %, No auto increment, Sample volume 20 µl, Reporter Dye: FAM, VIC, Quencher Dye: MGB, Passive Reference: ROX. Threshold: 0.12. Standard-mode run settings may be employed on non-fast machines, but optimal conditions may be subject to further modification.

8. When the reaction has completed, using the machine's software, label wells with appropriate identifiers and assign wells as plasmids or samples. Enter the appropriate absolute values for the plasmid wells to be used as standards. Proceed with automated analysis of the data to generate absolute quantification values for *ABL1* and *BCR-ABL1* for each well (*see* Figs. 1 and 2 and **Note 17**).

9. Calculate the average *ABL1* and *BCR-ABL1* value for each sample, excluding outliers (the difference between the highest and lowest replicates should be <0.5 for quantification cycle (C_t) values up to 30; the difference between the highest and lowest replicates should be <1.0 for C_t values between 30.1 and 33; the difference between the highest and lowest replicates should be <1.5 for C_t values between 33.1 and 37); above 37.1 C_t, replicates may show considerable variation and are outside the range of accurate quantification [9] (*see* Fig. 2).

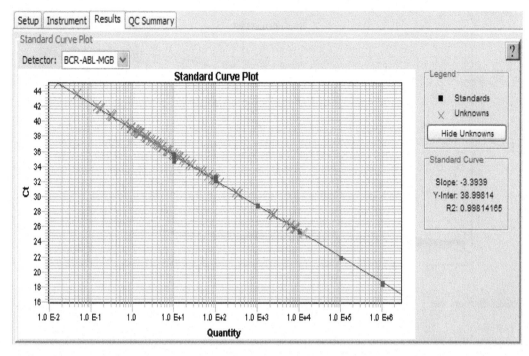

Fig. 1 Standard curve obtained from an ABI 7900HT machine, showing the C_t value (*y*-axis) plotted against the log of the BCR-ABL1 quantity (*x*-axis). The plasmid (standards) data points are represented by *filled squares* and the sample data points (unknowns) are represented by *red crosses*. Slope values should be between −3.2 and −3.6, R_2 should be greater than 0.98

10. Samples with positivity in only one of the triplicates should be considered as "borderline" between positive and undetectable.

11. If the *ABL1* value is less than 10,000 interpret the results with caution as this indicates poor sample quality and quantification is likely to be inaccurate and there is an increased risk of a false negative result (*see* **Note 18**).

12. A percentage ratio of *BCR-ABL1* for each sample is calculated as: $100 \times$ (average *BCR-ABL1* value/average *ABL1* value).

4   Notes

1. The isolation of total white cells and lysis process must occur within 72 h from collection of blood for reliable quantification, especially for follow up samples. For pretreatment, samples can be processed up to 4–5 days from collection. However, a qualitative rather than a quantitative test is likely to be more successful under these circumstances. EDTA is the preferred anticoagulant; lithium heparin is not recommended, because of possible interference with downstream PCR.

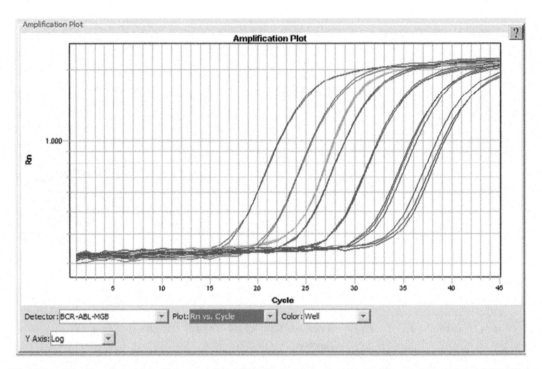

Fig. 2 Amplification plot obtained from an ABI 7900HT machine, showing the fluorescence of the BCR-ABL1 reporter dye (divided by that of a passive reference dye) given as Rn on the *y*-axis, plotted against the cycle number of the reaction (*x*-axis). The plasmid triplicates (at 10, 100, 1,000, 10,000, 100,000, and 100,000 molecules) are in *blue* and *purple*. A patient's sample trace (also in triplicate) is shown in *green*, the average BCR-ABL1 quantity of this triplicate was calculated at 13,293 molecules. Note the increased variance in plasmid triplicates at the lower quantities, exemplifying the increasing range of acceptance for triplicate values at these C_t levels

2. In CML total white cell isolation from whole blood is recommended (selective isolation of nucleated cells can lead to skewed results).

3. Physical control of contamination is critical throughout the whole procedure to reduce the risk of false positive results. At each step tubes should be opened and closed one at a time, samples should be handled slowly and carefully to avoid spills/ aerosols and gloves changed regularly to avoid carry over.

4. The loss of viscosity indicates that both high-molecular-weight genomic DNA has been degraded and that the sample has been adequately homogenized.

5. An alternative method of homogenization and disruption is to add a sterile 5 m stainless steel bead, seal the tube and transfer to the tissue homogenizer (e.g., Qiagen TissueLyserII) and shake at 20 Hz for 12 min.

6. A tube containing 1 ml of the RLT and beta-mercaptoethanol stock used should also be processed to monitor for contamination of reagents.

7. RNA extraction using Qiagen spin columns can be automated using a QIAcube sample preparation machine (Qiagen). Follow the RNeasy animal tissue and cells standard protocol with the elution volume adjusted to 60 μl.

8. The final volume of eluate is approximately 55 μl as some water is lost during the spin process through the column.

9. The quantity of RNA recovered can vary between samples but is within the acceptable range for the reverse transcriptase reaction.

10. The cDNA reagent mix can be prepared in advance (without the enzymes) and stored in aliquots at –20 °C, choose volumes convenient for the expected number of samples to be tested as this will vary depending on sample workload.

11. Enzymes are added immediately prior to use and are mixed by gentle pipetting. Calculate volumes for 10 % more than the actual number of samples to compensate for loss during pipetting.

12. An additional tube containing 55 μl of dH_2O and 45 μl of cDNA buffer and enzyme per batch can be added to monitor for contamination of reagents used in the cDNA synthesis process.

13. No-template PCR controls must be included, each containing 3 μl of ddH_2O (plus the 10 μl of Taqman Fast Advanced Master Mix and 7 μl of primer and probe mixture).

14. The primer and probe mixture can be prepared in advance and stored in aliquots at –20 °C. Choose volumes convenient for the expected number of samples.

15. Fast Advanced Master Mix and primer–probe mixes can be mixed immediately prior to use. Use of an 8 channel pipette and a reagent trough improves speed of aliquoting the mixture to the plate.

16. Use of an electronic pipette to aliquot the 3 μl volumes reduces variation between wells and improves set up time. Pipetting into 96-well plates can be tiring, to reduce the chance of error, pipette the samples as a 3 μl droplet onto the rear side of the wells, this allows a quick visual check of which wells have been dealt with.

17. Longitudinal monitoring of either plasmid C_t values or ratios of control samples can be conducted to identify any drift or significant changes in results.

18. Repeating the process from extraction of the RLT sample may improve the result.

References

1. Barnes DJ, Melo JV (2002) Cytogenetic and molecular genetic aspects of chronic myeloid leukaemia. Acta Haematol 108:180–202

2. Deininger MW, Goldman JM, Melo JV (2000) The molecular biology of chronic myeloid leukemia. Blood 96:3343–3356

3. Marin D, Ibrahim AR, Lucas C et al (2012) Assessment of BCR-ABL1 transcript levels at 3 months is the only requirement for predicting outcome for patients with chronic myeloid leukemia treated with tyrosine kinase inhibitors. J Clin Oncol 30:232–238

4. Ibrahim AR, Eliasson L, Apperley JF et al (2011) Poor adherence is the main reason for loss of CCyR and imatinib failure for chronic myeloid leukemia patients on long-term therapy. Blood 117:3733–3736

5. Faderl S, Talpaz M, Estrov Z et al (1999) The biology of chronic myeloid leukemia. N Engl J Med 341:164–172

6. Fujisawa S, Nakamura S, Naito K et al (2008) A variant transcript, e1a3, of the minor BCR-ABL fusion gene in acute lymphoblastic leukemia: case report and review of the literature. Int J Hematol 87:184–188

7. Foroni L, Gerrard G, Nna E et al (2009) Technical aspects and clinical applications of measuring BCR-ABL1 transcripts number in chronic myeloid leukemia. Am J Hematol 84:517–522

8. Beillard E, Pallisgaard N, van der Velden VHJ et al (2003) Evaluation of candidate control genes for diagnosis and residual disease detection in leukemic patients using 'real-time' quantitative reverse-transcriptase polymerase chain reaction (RQ-PCR)—a Europe against cancer program. Leukemia 17: 2474–2486

9. Gabert J, Beillard E, van der Velden VHJ et al (2003) Standardization and quality control studies of 'real-time' quantitative reverse transcriptase polymerase chain reaction of fusion gene transcripts for residual disease detection in leukemia—a Europe against cancer program. Leukemia 17:2318–2357

10. Foroni L, Wilson G, Gerrard G et al (2011) Guidelines for the measurement of BCR-ABL1 transcripts in chronic myeloid leukaemia. Br J Haematol 153:179–190

11. Gerrard G, Mudge K, Foskett P et al (2012) Fast-mode duplex qPCR for BCR-ABL1 molecular monitoring: innovation, automation, and harmonization. Am J Hematol 87: 717–720

Chapter 12

A Reliable Assay for Rapidly Defining Transplacental Metastasis Using Quantitative PCR

Samantha Mascelli

Abstract

To choose the most appropriate treatment for children affected by a transplacental metastasis, it is crucial to ascertain the maternal origin of the tumor. Up-to-date conclusive diagnosis is generally achieved through fluorescence in situ hybridization or karyotyping analysis.

Herein, we report an alternative, reliable assay for rapidly defining vertical cancer transmission to the fetus by using quantitative polymerase chain reaction. Our assay indicates that quantification of the copy number of the sex chromosomes by specific short tandem repeats markers, in genomic DNA purified from the tumor biopsy cells, could be used to correctly evaluate transplacental metastasis events.

Key words Transplacental metastasis events, qPCR, Molecular diagnosis

1 Introduction

Real-time quantitative PCR is used in a variety of fields, such as clinical diagnosis, molecular research, and forensics studies, to detect the presence of copy number changes in the genome or the viral/bacterial load in various body fluids, or to quantify the expression levels of specific genes [1]. Regardless of its application, qPCR is currently used to detect either DNA or RNA molecules. Recently, we developed a new qPCR application in an unusual clinical setting, that is, in a case of transplacental metastasis. In fact, we had to conclusively show the maternal origin of a rare temporal bone tumor in an infant. Cancer transmission to the fetus is uncommon, and very few reports have ever shown such an event [2], even though cancer during pregnancy is not an exceptional event (1 case per 1,000 live births) [3]. Biopsy can significantly contribute to the correct management of patients with transplacental metastases. Apart from a solid histologic diagnosis, this procedure allows us to obtain tissue for cytogenetic analysis [karyotyping or fluorescence in situ hybridization (FISH)], which will then allow us to correctly distinguish maternal cells from fetal cells. These methods require

Roberto Biassoni and Alessandro Raso (eds.), *Quantitative Real-Time PCR: Methods and Protocols*, Methods in Molecular Biology, vol. 1160, DOI 10.1007/978-1-4939-0733-5_12, © Springer Science+Business Media New York 2014

an adequate amount of good-quality tissue samples, and they must be performed by a skilled cytogeneticist. We therefore successfully developed a quick and simple screening method to show the maternal origin of tumors. It is based on the evaluation of selected markers mapping on the sex chromosomes of the tumor cells, and may be used in cases of vertical transmission to a male fetus. This approach allows to correctly evaluate the maternal origin of tumor cells using qPCR. It may be added to the analyses that are commonly used for early sex determination by detection of fetal DNA in maternal plasma or for sex determination and genotyping both in genetic paleontology and forensic analysis.

2 Materials

Prepare all solutions using ultrapure water (prepared by purifying deionized water to attain a sensitivity of 18 MΩ/cm at 25 °C) and analytical grade reagents.

Prepare and store all reagents at room temperature (unless indicated otherwise).

Prepare and store all the reagents at 4 °C unless otherwise indicated.

Wear gloves to prepare all reagents.

2.1 Genomic and Somatic DNA Extraction

1. Extract genomic DNA samples from peripheral lymphocytes by using GenElute Mammalian genomic DNA miniprep kit (Sigma-Aldrich).

2. Isolate somatic DNA samples from frozen tumor slides (*see* **Notes 1** and **2**) by using PureLink mini columns (Invitrogen, Carlsbad, CA).

3. Microcentrifuge and vortex for mixing preps.

4. Elution buffer: 10 mM Tris–HCl, 0.5 mM EDTA, pH 9.0.

5. Quantify DNA samples with Nanodrop spectrophotometric analyzer (Celbio, Milan, Italy).

2.2 Amplification of Sex Chromosome-Specific Regions

1. The primers and probes sequences and their working concentration are shown in Table 1 (*see* **Note 3**).

2. Sex chromosome STR markers are: DYS14 mapping on chromosome Y, DXS6803 and GATA165B12 on chromosome X, and telomerase loci as the control [4–6] (*see* **Note 4**).

3. Real-time PCR mastermix (Invitrogen, Carlsbad, CA).

4. ABI 7500 Instrument (Applied BioSystems).

5. MicroAmp™ Optical 96-Well Reaction Plate (Applied BioSystems) (*see* **Note 5**).

Table 1
Primers and working concentration

Primers	Sequences 5′–3′	Concentrations (nM)
DYS14 forward	GGCCAATGTTGTATCCTTCT	200
DYS14 reverse	CCCATCGGTCACTTACACTT	100
DXS6803 forward	GAAATGTGCTTTGACAGGAA	450
DXS6803 reverse	CAAAAAGGGACATATGCTACTT	900
Telomere forward	GTGAACCTCGTAAGTTTATGCAA	50
Telomere reverse	GCACACGTGGCTTTTCG	100
GATA165B12 forward	TATGTATCATCAATCATCTATCCG	900
GATA165B12 reverse	TTAAAATCATTTTCACTGTGTATGC	300

6. MicroAmp™ Optical Adhesive Film (Applied BioSystems).

7. Nuclease-free water.

8. Microcentrifuge and vortex for mixing preps.

9. Tubes RNase, DNase, DNA and PCR inhibitors free.

10. MicroPipettes (single and multichannel).

11. Filter tips (RNase, DNase, DNA and PCR inhibitors free).

12. Disposable gloves.

13. Bio containment hoods (*see* **Note 6**).

3   Methods

3.1   DNA Samples Extraction

Carry out all procedures at room temperature (RT) unless otherwise specified.

Fifty-hundred microliters of the whole blood or the pool of slides of frozen tissue is suspended in the Suspension Solution, as indicated by the reference kit. DNA extraction is carried out according to the manufacturer's protocol, eluting the DNA template in 60 µl (for each 0.5 ml of original sample) of Elution Buffer (*see* **Note 7**).

3.2   Primers Optimization

A preliminary thermodynamic analysis of the published primers of the selected markers was performed [5, 6]. Thus, both the primer concentrations and the sequence length were optimized until comparative standard curves (SCs) were reached using several different types of male genomic DNA (Table 1 and Fig. 1). The greatest efficiency was reached by using DXS6803, which generated an SC that was comparable to that of the DYS14 marker, whereas the GATA165B12 did not achieve comparable amplification efficiency (Fig. 1).

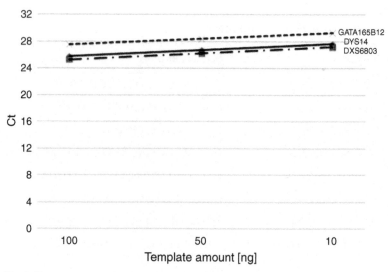

Fig. 1 Standard curves for the selected short tandem repeats (STR) markers using several samples of peripheral male blood DNA: DXS14 and DXS6803 qPCR assays run in singleplex. DYS14 Slope = −3.3354 with R^2 = 0.90; DXS6803 Slope = −3.1233 with R^2 = 0.99. GATA165B12 Slope = −2.951 with R^2 = 0.95. $\Delta Ct^{(DYS14-DXS6803)}$ = 0.4

3.3 Standard Curve (SCs) Preparation

Both the sensitivity and reproducibility of each assay were ensured by generating SCs using several peripheral blood DNA samples at various concentrations (Fig. 1). Robustness was evaluated by using a mixture of male and female DNA, ranging from 100 % XY DNA to 100 % XX DNA, stepping increasing by 25 % fractions (Fig. 2a). The normalized fluorescent signal (ΔRn) was automatically calculated by an algorithm that normalizes the reporter emission signal. The threshold value that was applied to the algorithm generating the threshold cycle (Ct) was set at 0.05 in all the experiments.

3.4 qPCR Procedure

The relative quantification of each short tandem repeat (STR) was performed according to the comparative method (ΔCt, Applied Biosystems User Bulletin no. 2P/N4303859). Amplifications were carried out in singleplex runs on 25 μl 20 ng of DNA and using Platinum QPCRSYBR-GREEN SuperMix-UDG (Invitrogen) on the ABI PRISM 7500 HT Sequence Detection System (Applied Biosystems). Cycling conditions included degradation of preamplified templates for 2 min at 50 °C, followed by 2 min of denaturation at 95 °C, 32 cycles of denaturation at 95 °C for 20 s, and annealing/extension at 55 °C for 35 s, followed by the dissociation stage.

3.5 Assay Validation

The relative quantifications of sex chromosomes were carried out by using DYS14 as the target, normalized to the DXS6803 as the calibrator using a range of male DNA concentrations ranging from 10 to 100 ng per reaction. The reproducibility of the calibration

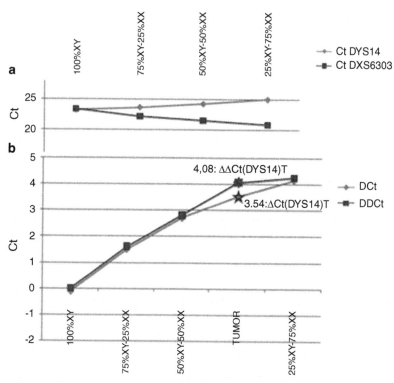

Fig. 2 Evaluation of DYS14 and DXS6303 on a mixture of male and female DNA, ranging from 100 % XY DNA to 100 % XX DNA stepping with 25 % fractions. (**a**) Cts only show a correct identification of XY–XX ratio: 50 % XY–50 % XX DNA (the theoretical ratio is 1:4), 100 % female DNA DYS14 is not amplified. (**b**) Tumor $\Delta Ct^{(DYS14)}$ and $\Delta\Delta Ct$. Maternal origin of the tumor is assessed by $\Delta Ct^{(DYS14)} = 3.54$ and $\Delta\Delta Ct^{(DYS14)} = 4.08$ corresponding to at least 60 % of XX cells

curve was analyzed by evaluating the slope and the correlation coefficient of the curve. DYS14 slope $= -3.3354$ with $R^2 = 0.99$; DXS6303 slope $= -3.1233$ with $R^2 = 0.99$. The qPCR efficiencies were calculated by the equation: $E = 10^{[-1/\text{slope}]}$ and the difference between the efficiencies ($E^{DYS14} - E^{DXS6803}$) was <0.1, indicating that the data could be compared [7]. The ΔCt ($Ct^{DYS14} - Ct^{DXS6803}$) indicated the same presence of sex chromosomes X and Y on male genomic DNA, as expected (Fig. 1). To determine the sensitivity of the method, samples containing various proportions of a mixture of male and female DNA were analyzed. The Ct target of DYS14 was normalized by the Ct of the DXS6303 by the equation $\Delta Ct^{(DSY14)} = Ct^{(DYS14)} - Ct^{(DXS6803)}$. The $\Delta\Delta Ct^{(DSY14)}$ for each DNA concentration mixture was calculated using the $\Delta Ct^{(DSY14)}$ at 100 % XY as the calibrator by the equation $\Delta\Delta Ct_X = \Delta Ct_X^{(DSY14)} - \Delta Ct_{100\%XY}^{(DYS14)}$. The results correctly identified the XY–XX ratio. In the absence of the Y chromosome (i.e., 100 % female DNA), DYS14 was not amplified, whereas with 50 % XY–50 % XX DNA (theoretical ratio 1:4), the ΔCt (DYS14) value was 2.71 and the DDCt (DYS14) value was 3.02 (Fig. 2b).

3.6 Detection of Chromosome Set in the Patient's Biopsy

By using qPCR on DYS14 and DXS6803 (*see* **Note 8**), the same amount of the two markers (Y chromosome and X chromosome) was found in the DNA extracted from the peripheral blood, as expected, $\Delta Ct_{blood(DYS14)} = -0.4/-0.5$, whereas the DNA from the tumor biopsy showed $\Delta Ct_{tumor}(DYS14) = 3.54$, indicating fewer Y chromosome markers than X chromosome markers (Fig. 2). The $\Delta\Delta Ct$ calculated as $\Delta Ct_{tumor\ (DYS14)} - \Delta Ct_{blood\ (DYS14)}$ was 4.08 (Fig. 2b), thus indicating that at least 60 % of cells had the XX chromosome set. The results did not change even when the Cts of telomerase loci were used as a control to generate $\Delta\Delta CtDSY14$. The result was confirmed by FISH.

4 Notes

1. The frozen tumor biopsy are processed by drawing 6 mm sections that are stained with hematoxylin and eosin, and reviewed by the pathologist to make sure that the tumor cell content is above 80 %.

2. Other frozen sections are collected in cryovials for the following DNA extraction.

3. To obtain the greatest efficiency of the systems, the published primer sequences were modified using Primer Express (PE Applied Biosystem, Foster City, CA), Oligo 4.1 (National Biosciences Inc., Plymouth, MN), and PrimerPy v0.97 (a GUI utility for Q-PCR primer design software). The best thermodynamically performing amplification was reached by varying the reaction conditions and the primer concentrations (Table 1).

4. Amplified sex chromosome-specific regions, such as short tandem repeats (STR) loci, were used to design a system that could detect the possible presence of maternal cells within the biopsy sample of the infant's tumor. Among the possible sex chromosome STR markers, we selected the most often used and efficient ones reported in the literature.

5. The covers primarily require the application of pressure by the user to ensure a tight, evaporation-free seal. Improper peeling of the cover may result in haziness but does not affect results.

6. To prevent DNA contamination of laboratory and samples.

7. DNA samples can be stored at 4 °C when planning a real-time PCR run within a week, otherwise at −20 °C for longer storage.

8. The strategy for assessing the maternal origin of the tumor was founded on the difference in sex chromosomes between the mother and son. We searched for the presence of XX cells within the tumor mass using the peripheral blood DNA as the calibrator.

References

1. VanGuilder HD, Vrana KE, Freeman WM (2008) Twenty-five years of quantitative PCR for gene expression analysis. Biotechniques 44: 619–626

2. Valasek MA, Repa JJ (2005) The power of real-time PCR. Adv Physiol Educ 29: 151–159

3. Alexander A, Samlowski WE, Grossman D et al (2003) Metastatic melanoma in pregnancy: risk of transplacental metastases in the infant. J Clin Oncol 21:2179–2186

4. Picchiassi E, Coata G, Fanetti A et al (2008) The best approach for early prediction of fetal gender by using free fetal DNA from maternal plasma. Prenat Diagn 28:525–530

5. Zhong XY, Holzgreve W, Hahn S (2006) Direct quantification of fetal cells in maternal blood by real-time PCR. Prenat Diagn 26:850–854

6. Vecchione G, Tomaiuolo M, Sarno M et al (2008) Fetal sex identification in maternal plasma by means of short tandem repeats on chromosome X. Ann N Y Acad Sci 1137: 148–156

7. Pfaffl MW (2001) A new mathematical model for relative quantification in real-time RT-PCR. Nucleic Acids Res 29:e45

Chapter 13

Circulating Cell-Free DNA in Cancer

Pamela Pinzani, Francesca Salvianti, Claudio Orlando, and Mario Pazzagli

Abstract

This papers deals with the preanalytical and analytical phase of cell-free DNA analysis, highlighting some criticism on sample collection and extraction. We describe a method to accurately quantify total cfDNA in plasma and our particular approach to the measurement of tumor deriving cfDNA.

Key words Cell-free DNA, qPCR, *BRAFV600E*, Circulating markers, Plasma

1 Introduction

The discovery of nucleic acids circulating in the blood (CNA) has influenced the scientific scenario of the last decade and today represents one of the most interesting topics in the field of oncology [1]. The generic term CNA identifies not only DNA, but also RNA isolated from plasma, serum, and other body fluids such as urine or lymph [2].

Cell-free DNA (cfDNA) has been firstly reported in 1948 by Mandel and Metais [3], but at the time no association with disease was hypothesized. Only 30 years later, in 1977, Leon et al. [4] found cfDNA in plasma of patients affected by lung cancer.

The origin of cfDNA is still under debate, but three main hypotheses are supported. They are supposed to derive from cell apoptosis [5, 6] or alternatively from cell necrosis, but some authors reported the possibility of an active release from cells [6–8]. The presence of DNA circulating freely in the bloodstream of healthy subjects can be related to the lysis of activated lymphocytes and of other nucleated cells or to their active secretion of nucleic acids. In patients affected by neoplastic diseases it is supposed that normal and cancer cell can:

1. Detach from the tumor mass and undergo necrosis or apoptosis.

2. Actively release nucleic acids in the blood flow [9].

Roberto Biassoni and Alessandro Raso (eds.), *Quantitative Real-Time PCR: Methods and Protocols*, Methods in Molecular Biology, vol. 1160, DOI 10.1007/978-1-4939-0733-5_13, © Springer Science+Business Media New York 2014

Although the physical and chemical properties of circulating nucleic acids and the mechanisms of their release into the circulation are not yet well understood, their potential applications are of great interest [10].

In particular, cfDNA, due to its stability and relative abundance, seems to represent a good tool for clinical applications, also providing a non-invasive surrogate for molecular analysis in cancer and pre-cancer patients. cfDNA can be evaluated quantitatively as total plasma DNA concentration. However, an increased plasma DNA level is not only detectable in patients with cancer or with premalignant stage of the disease, but also as a consequence of inflammation, trauma and in elderly patients suffering from acute or chronic illnesses [11]. Thus, the availability of innovative techniques able to detect the presence of cancer-associated genetic or epigenetic alterations, even in a low amount of DNA, makes cfDNA as an amenable and a more specific tumor marker.

In addition the identification of tumor-derived cfDNA and its molecular characterization could represent a key tool for gaining results that may allow a better classification of the different subsets of cancer patients with different prognosis and, more specifically, the identification of the cellular profiles underlying aggressiveness or responsiveness to therapy [12].

Genetic or epigenetic alterations have been previously investigated in cfDNA of patients with tumors [2]. Preliminary data arising from the recent literature suggest promising areas of application for this type of biomarkers. It will be crucial to determine which combination of genetic alterations may carry the best prognostic or diagnostic value in oncologic patients.

Most of the previous studies have mainly focused on the qualitative evaluation of a single molecular marker (i.e., presence versus absence of one identifiable alteration in cfDNA) instead of the quantitative analysis of a single marker or a combination of molecular markers. The search for new clinical and prognostic indicators needs an equally significant methodological effort. In this respect, qPCR analysis represents one of the suitable analytical tools for the quantitative analysis of molecular targets present in such low amount in plasma samples. With regard to the detection of tumor related genetic variants, the major disadvantage of qPCR is represented by its ability to identify only predefined mutations and by the requirement of specific assay formats conferring the ability to discriminate the mutant molecular variant differing for only one nucleotide from the wild type sequence. On the other side, the quantitative approach, the high sensitivity and specificity affordable make this technique particularly appealing for the analysis of cfDNA. A careful evaluation of the common parameters of sensitivity, specificity and accuracy for each plasma DNA target is thus required.

Moreover, the preanalytical phase still presents critical aspects for these applications due to the need for a standardized protocol for sample collection together with a suitable technique of DNA extraction from plasma.

The achievement of quantitative measurements is another critical issue. Different studies are based on the choice of several genes, on different amplicon size and measurement protocols. All these aspects contribute cumulatively to impair the comparison of data from different studies and to draw definitive conclusions about the real impact of the diagnostic parameter.

The aim of this manuscript is precisely to provide an evaluation of the critical aspects related to cfDNA, focusing on technical problems, so as to provide helpful advice on free plasma DNA testing.

The topic will deal with a sample collection protocol and the description of a modified procedure for DNA purification from plasma.

Once optimized all the preanalytical aspects, a description of a method used for total cfDNA measurement will follow together with our particular approach to the measurement of *BRAFV600E* gene variant in plasma.

2   Materials

2.1   Sample Collection

1. EDTA tubes.
2. 1.5 ml tubes.
3. Micropipets.
4. 1 ml pipet tips.
5. Refrigerated centrifuge for 15 ml tubes.
6. Refrigerated microcentrifuge.
7. Disposable gloves.

2.2   cfDNA Extraction

1. QIAamp DSP virus Kit:
   - Refer to the QIAamp DSP virus Kit Handbook for a detailed description of the reagents and the manufacturer's protocol.
   - Prepare buffer AW1 and AW2 by adding the appropriate volume (respectively 25 ml and 30 ml) of ethanol (96–100 %).
   - Resuspend protease in the appropriate solvent.
   - In order to increase the recovery rate of plasma DNA add carrier RNA to buffer AL at a final concentration of 11.2 µg/ml.
2. Ethanol (96–100 %).
3. Pipets and pipet tips.

4. Disposable gloves.

5. Heating block.

6. Microcentrifuge.

7. Vortexer.

2.3 qPCR

1. Spectrophotometer (e.g., NanoDrop).

2. qPCR instrument (e.g., 7900HT, Life Technologies).

3. Plasticware specific for the qPCR instrument.

4. Pipets and pipet tips.

5. Primers and probes (*see* the paragraphs below for details).

6. QuantiTect Probe PCR Master Mix (Qiagen).

7. PCR grade water.

3 Methods

3.1 Sample Collection

Sample collection represents a critical issue when dealing with cfDNA. In fact, when blood is drawn according to standard methods, two processes are supposed to occur over time (hours to days) if the blood is not immediately processed for diagnostics testing: nucleic acids may degrade due to the presence of nucleases in whole blood or plasma fraction, respectively. In addition, nucleated blood cells will die and disintegrate over time, releasing comparatively large amounts of genomic DNA into the plasma fraction, thus reducing the fractional concentration of the cfDNA originally present in the samples and, most importantly, diluting potential diagnostic targets such as rare tumor-derived DNA fragments.

The role of the delay in blood processing and of the storage temperature on the amount of DNA extracted from plasma has been demonstrated [13] showing an increase 24 h after venipuncture. On the contrary, anticoagulants seem not to interfere on the quantity of the recovered DNA from plasma, but EDTA shows a stabilizing effect on blood during the time between sample draw and processing, both at room temperature and at 4 °C [14].

In order to get rid of contaminating DNA deriving from cells, both filtration [15] and repeated centrifugations [16–18] at low and high speed were reported, demonstrating that no release of circulating nucleic acid was induced from blood cells even at maximum centrifugation speed.

In our lab the following procedure is adopted for plasma sample collection:

1. Samples are collected in EDTA tubes and are received by the lab within 1 h from blood draw.

2. The blood samples are submitted to a first centrifugation step at $1,600 \times g$, 4 °C for 10 min.

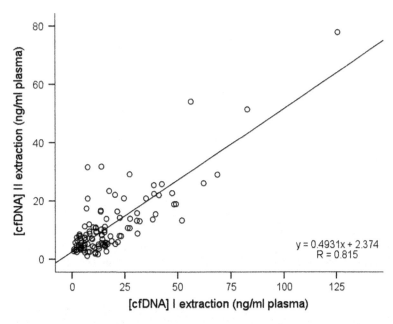

Fig. 1 Effect of storage at −80 °C of plasma samples on the quantity of cfDNA. We investigated the effect of plasma storage at −80 °C on the total amount of cfDNA; 122 plasma samples collected following the previously reported procedure were processed twice with a time lapse ranging from 5 to 21 months. The first extraction was performed within 1 month from blood collection. The correlation between the quantitative measurement of the total cfDNA quantity after the first (*X* axis: *[cfDNA] I extraction*) and second (*Y* axis: *[cfDNA] II extraction*) extraction of two different aliquots of the same plasma sample resulted in a statistically significant relationship between the two sets of measurements

3. Plasma is transferred to new 1.5 ml tubes.

4. A second centrifugation is performed at maximum speed, at 4 °C, for 10 min. Pellets (if any) are discarded.

5. Plasma is split into one-extraction-aliquots of 500 μl each. Plasma is maintained at −80 °C until extracted.

Regarding the stability of CNA in the frozen samples, some authors showed that plasma can be conserved frozen for years (at least 2 for RNA and 6 for DNA) [19–21] at −70 or −20 °C without affecting CNA concentration, while other authors reported a decay of 30 % in DNA from stored plasma [22] (*see* **Note 1**) (Fig. 1).

3.2 Cell-Free DNA Extraction from Plasma

In our lab we adopted the QIAamp DSP virus Kit, since no specific kits for cfDNA were available when we began studying cfDNA. Initially we performed a comparison between QIAamp DSP virus Kit and another extraction system optimized for blood (*see* **Note 2** and Fig. 2a). New methods (*see* **Note 3**) now available for the extraction of cfDNA have been shown to produce similar results (Fig. 2b).

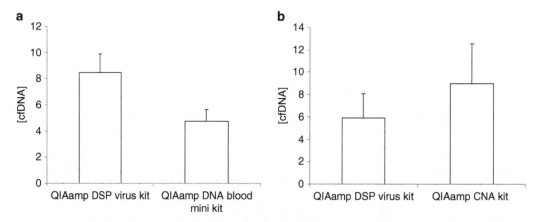

Fig. 2 Comparison between two different methods of extraction. (**a**) Comparison between QIAamp DSP virus kit and QIAamp DNA blood mini kit on 26 plasma samples (*t*-test for paired samples $p = 0.004$). (**b**) Comparison between QIAamp DSP virus and QIAamp CNA kit on 21 plasma samples

The QIAamp DSP virus Kit procedure was modified to meet the laboratory requirements in order to use centrifugation instead of a vacuum system. The protocol is reported below.

1. Pipet 75 μl Protease into a lysis tube.

2. Add 500 μl of plasma.

3. Add 500 μl lysis Buffer AL (containing 11.2 μg/ml of carrier RNA). Close the cap and mix by pulse-vortexing for 15 s.

4. Optional: add 2 μl RNAse A and mix by vortexing briefly.

5. Incubate at 56 °C for 15 min.

6. Centrifuge the lysis tube for at least 5 s at maximum speed to remove drops from the inside of the lid.

7. Change gloves and open carefully the lysis tube.

8. Add 600 μl ethanol (96–100 %) to the sample, close the lid, and mix thoroughly by pulse-vortexing for 15 s. Incubate for 5 min at room temperature (15–25 °C).

9. Centrifuge the lysis tube for at least 5 s at maximum speed to remove drops from the inside of the lid.

10. Apply 600 μl lysate onto the QIAamp MinElute column
    - Centrifuge at $13,000 \times g$ for 1 min.
    - Discard the flow-through.
    - Repeat this step until the whole lysate has been loaded onto the column.

11. Add 600 μl Buffer AW1
    - Centrifuge at $13,000 \times g$ for 1 min.
    - Discard the flow-through.

12. Add 750 µl Buffer AW2
    - Centrifuge at $13,000 \times g$ for 3 min.
    - Discard the flow-through.
13. Add 750 µl ethanol (96–100 %)
    - Centrifuge at $13,000 \times g$ for 3 min.
    - Discard the flow-through.
14. Centrifuge at $13,000 \times g$ for 1 min
    - Discard the collection tube.
15. Insert the column into a new 1.5 ml tube.
16. Open the lid and incubate then at 56 °C for 3 min to dry the membrane.
17. Place the QIAamp MinElute column in an elution tube (ET)
    - Apply 20 µl Buffer AVE to the center of the membrane.
    - Close the lid and incubate at room temperature for 5 min.
    - Centrifuge at full speed for 1 min.

3.3 Quantification of cfDNA by qPCR

1. Assay design. Absolute quantification of cfDNA is obtained by amplifying a target sequence of the single copy gene *APP* (Amyloid Precursor protein, chr.4q11-q13) by qPCR by means of the following primers and hydrolysis probe:

    Forward Primer: 5′-TCAGGTTGACGCCGCTGT-3′.

    Reverse Primer: 5′-TTCGTAGCCGTTCTGCTGC-3′.

    Hydrolysis Probe: 5′-FAM-ACCCCAGAGGAGCGCCAC CTG-TAMRA-3′.

2. Quantify a genomic DNA at the spectrophotometer and prepare five ten-fold serial dilutions ranging from 10^5 to 10 pg/µl (Fig. 3a).

3. Include the five dilutions of the standard curve and suitable control samples in each run.

4. Include No-Template Control (NTC) reactions.

5. Prepare a PCR reaction mix (11.5 µl per sample replicate) according to the following scheme:
    - 1× QuantiTect Probe PCR Master Mix (Qiagen).
    - 300 nM forward primer.
    - 300 nM reverse primer.
    - 200 nM probe.
    - PCR grade water.

6. Add 1 µl of cfDNA sample.

7. Test each sample in duplicate.

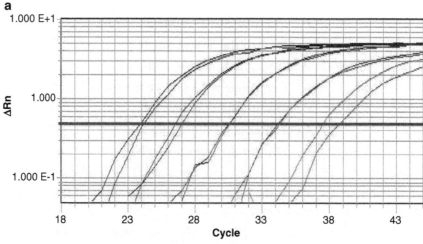

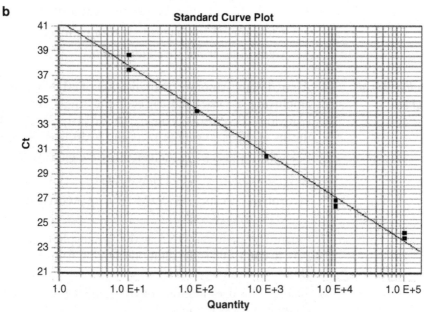

Fig. 3 qPCR assay method for cfDNA concentration in plasma. (**a**) *APP* gene amplification plots of reference samples containing known quantities of genomic DNA. (**b**) Standard curve obtained by linear regression analysis of the reference samples ($y = -3.5x + 41.4$ $R^2 = 0.99$)

8. Run the PCR reaction in a real time PCR instrument according to the following thermal profile: 95 °C for 10 min and 45 cycles of PCR (95 °C for 15 s, 60 °C for 60 s).

9. The concentration of unknown samples will be obtained by interpolating data on the reference curve (Fig. 3b).

3.4 Measurement of Plasma Tumor DNA

In affected patients, DNA concentration in plasma can be influenced by tumor stage, size, and location [23]. However, these values may also be altered in patients with various diseases (such as trauma, stroke, burns, sepsis, and autoimmune diseases), thus limiting their value for the diagnosis of cancer [24].

Therefore quantitative analyses limited to cfDNA concentration could not provide the expected clinical specificity, unless combined with qualitative alterations of DNA, such as mutations, loss of heterozygosity (LOH), microsatellite instability, and epigenetic changes [23]. Dealing with point mutations, their scarcity in plasma if compared to wild-type sequences and the high sensitivity requested for their detection represent the two main obstacles for immediate clinical application. The detection of these biomarkers implies the assessment of the optimal analytical conditions in order to achieve the highest sensitivity with the maximum specificity. Here we report an example of an allele specific qPCR assay designed and optimized for the detection of *BRAFV600E* mutated alleles in plasma [25]. *BRAF* somatic mutations have been reported in a wide range of human cancers, with the highest frequency in melanoma and thyroid cancer [26]. The possibility of reliably detecting *BRAF*-mutated DNA in plasma [27–29] could have several important clinical applications in the short- and long-term follow-up of cancer patients particularly referring to melanoma and papillary thyroid carcinoma.

3.5 Quantification of BRAFV600E Mutated Alleles by qPCR

1. Assay design. Specificity for the *BRAFV600E* mutated allele is obtained by means of the forward primer and a LNA (Locked Nucleic Acid) probe (Sigma, USA), while the reverse primer recognizes both wild type and mutated sequences. The sequence of the primers and probe are reported below:

   Forward   primer:
   5′-AAAATAGGTGATTTTGGTCTAGCTACAGA-3′.

   Probe: 5′-FAM-[+C]GAGA[+T]TT[+C][+T][+C]TG[+T] AG[+C]TBHQ1-3′.

   Reverse primer: 5′-GACAACTGTTCAAACTGATGG-3′.

2. Standard curve preparation. To obtain an absolute quantification of mutate alleles include a reference curve in each run (Fig. 4a).

   The standard curve for *BRAFV600E* consists of five dilutions (100, 50, 20, 10, and 1 % mutated alleles) obtained by mixing DNA from a cell line homozygous for the mutation (i.e., human melanoma cell line SKMEL28) and a wild type cell line (i.e., human breast cancer cell line MCF7). The total amount of each standard DNA is 0.5 ng.

3. Include the five dilutions of the standard curve and the wild type control sample in each run.

4. Include NTC reactions.

5. Prepare a PCR reaction mix with the following final concentrations:

   • 1× Quantitect® Probe PCR Master Mix (Qiagen).

   • 200 nM forward primer.

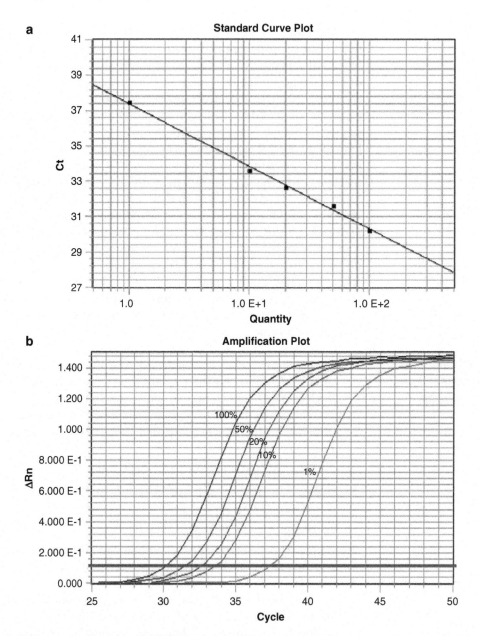

Fig. 4 qPCR assay method for *BRAFV600E*. (**a**) Amplification plots of reference samples containing known percentages (100, 50, 20, 10, and 1 %) of *BRAFV600E* mutated alleles. (**b**) Standard curve obtained by linear regression analysis of the reference samples ($y = -3.54x + 37.37$ $R^2 = 0.99$)

- 200 nM reverse primer.
- 200 nM LNA probe.
- PCR grade water.

6. Add 0.5 ng/reaction of cfDNA in a final reaction volume of 20 µl.

7. Test each sample in duplicate.

8. The thermal profile of the qPCR reaction is the following: 95 °C for 10 min and 50 cycles at 95 °C for 15 s and 64 °C for 1 min.

9. The percentage of *BRAFV600E* allele in unknown samples will be obtained by interpolating data on the reference curve (Fig. 4b).

The absolute concentration of *BRAFV600E* in terms of mutated DNA (ng/ml plasma) can be obtained by quantifying total DNA (*see* Subheading 3.3) and applying the following formula:
$$BRAFV600E \text{ (ng/ml plasma)} = (\% \ BRAFV600E) \times (\text{total DNA ng/ml plasma})/100.$$

4   Notes

1. We investigated the effect of plasma storage at −80 °C on the total amount of cfDNA; 122 plasma samples collected following the previously reported procedure were processed twice with a time lapse ranging from 5 to 21 months. Correlation between the quantitative measurement of the total cfDNA quantity after the first and second extraction of two different aliquots of the same plasma sample resulted in a statistically significant relationship between the two sets of measurements ($p < 0.001$; Fig. 1). Nonetheless a mean decrease of 38 % in total cfDNA could be detected: cfDNA mean value at the time of first extraction was 17.35 ± 1.63 ng/ml plasma, while after the second extraction resulted 10.93 ± 0.99 ng/ml plasma (*t*-Test for paired samples, $p < 0.001$). Within the observation period, we could not evidence any statistical difference in DNA decrement on the basis of the storage period at −80 °C.

2. The QIamp Blood Mini Kit is a method of extraction aimed at the recovery of DNA with a length of 20–30 Kb to 50 Kb or higher. At first, most of the studies on cfDNA have been performed using this kit since no specific reagent was available. The protocol was performed on plasma samples using a starting volume of 400 μl. The procedure was modified and the first step, corresponding to cell lysis and protein degradation, was performed by the use of anionic detergents and proteinase K with double reagent volumes compared to those specified by the manufacturer's protocol (QIamp Blood Mini Kit Handbook). After appropriate washing, the DNA bound to the column was eluted in a final volume of 100 μl. DNA samples were stored at a temperature of −80 °C.

The QIamp DSP Virus Kit is a method of nucleic acid extraction from 500 µl plasma. The elution was carried out in a volume of 20 µl. After the extraction the sample was stored at a temperature of −80 °C.

Comparison of the proposed methods confirmed a better performance for the DSP virus kit in comparison to the kit developed for blood processing (Fig. 2a).

3. QIamp Circulating Nucleic Acid Kit recently came to the market and it is specifically developed for CNA extraction. It is based, analogously to the previous methods, on the selective binding properties of silica-based membrane, but takes advantage of flexible plasma starting volumes (range 500 µl to 5 ml) and elution volumes (between 20 and 150 µl). DNAse/RNAse digestion can be applied to avoid interference by the non-targeted nucleic acid. The protocol was performed following the manufacturer's instructions.

No statistical difference in the quantity of cfDNA could be evidenced between DSP virus and CNA kit (Fig. 2b).

References

1. Tsang JC, Lo YM (2007) Circulating nucleic acids in plasma/serum. Pathology 39:197–207

2. Fleischhancker M, Schmidt B (2006) Circulating nucleic acids (CNAs) and cancer—a survey. Biochim Biophys Acta 1775:181–232

3. Mandel P, Metais P (1948) Les acides nucléiques du plasma sanguine chez l'homme. C R Acad Sci Paris 142:241–243

4. Leon SA, Shapiro B, Sklaroff DM et al (1977) Free DNA in the serum of cancer patients and the effect of therapy. Cancer Res 37:646–650

5. Georgiou CD, Patsoukis N, Papapostolou I (2005) Assay for the quantification of small-sized fragmented genomic DNA. Anal Biochem 339:223–230

6. Holdenrieder S, Stieber P (2004) Apoptotic markers in cancer. Clin Biochem 37:605–617

7. Nagata S (2005) DNA degradation in development and programmed cell death. Cell Death Differ 10:108–116

8. Jahr S, Hentze H, Englisch S et al (2001) DNA fragments in the blood plasma of cancer patients: quantitations and evidence for their origin from apoptotic and necrotic cells. Cancer Res 61:1659–1665

9. Gormally E, Caboux E, Vineis P et al (2007) Circulating free DNA in plasma or serum as biomarker of carcinogenesis: practical aspects and biological significance. Mutat Res 635:105–117

10. Gahan PB, Swaminathan R (2008) Circulating nucleic acids in plasma and serum. Recent developments. Ann N Y Acad Sci 1137:1–6

11. Gahan PB, Anker P, Stroun M (2008) Metabolic DNA as the origin of spontaneously released DNA? Ann N Y Acad Sci 1137:7–17

12. Stroun M, Anker P, Lyautey J et al (1987) Isolation and characterization of DNA from the plasma of cancer patients. Eur J Cancer Clin Oncol 23:707–712

13. Jung M, Klotzek S, Lewandowski M et al (2003) Changes in concentration of DNA in serum and plasma during storage of blood samples. Clin Chem 49:1028–1029

14. Lam NYL, Rainer TH, Chiu RWK et al (2004) EDTA is a better anticoagulant than heparin or citrate for delayed blood processing for plasma DNA analysis. Clin Chem 50:256–257

15. Chiang P, Beer DG, Wei W et al (1999) Detection of erbB-2 amplifications in tumors and sera from esophageal carcinoma patients. Clin Cancer Res 5:1381–1386

16. Chiu RWK, Poon LLM, Lau TK et al (2001) Effects of blood-processing protocols on fetal and total DNA quantification in maternal plasma. Clin Chem 47:1607–1613

17. Swinkels DW, Wiegerinck E, Steegers EA et al (2003) Effects of blood-processing protocols on cell-free DNA quantification in plasma. Clin Chem 49:525–526

18. Lui YY, Chik KW, Lo YM (2002) Does centrifugation cause the ex vivo release of DNA from blood cells? Clin Chem 48:2074–2076

19. Kopreski MS, Benko FA, Kwee C et al (1997) Detection of mutant K-ras DNA in plasma or serum of patients with colorectal cancer. Br J Cancer 76:1293–1299

20. Lee T, LeShane ES, Messerlian GM et al (2002) Down syndrome and cell-free fetal DNA in archived maternal serum. Am J Obstet Gynecol 187:1217–1221

21. Kopreski MS, Benko FA, Kwak LW et al (1999) Detection of tumor messenger RNA in the serum of patients with malignant melanoma. Clin Cancer Res 5:1961–1965

22. Sozzi G, Roz L, Conte D et al (2005) Effects of prolonged storage of whole plasma or isolated plasma DNA on the results of circulating DNA quantification assays. J Natl Cancer Inst 97:1848–1850

23. Jung K, Fleischhacker M, Rabien A (2010) Cell-free DNA in the blood as a solid tumor biomarker—a critical appraisal of the literature. Clin Chim Acta 411:1611–1624

24. van der Vaart M, Pretorius PJ (2008) Circulating DNA. Its origin and fluctuation. Ann N Y Acad Sci 1137:18–26

25. Pinzani P, Salvianti F, Cascella R et al (2010) Allele specific Taqman-based real-time PCR assay to quantify circulating BRAFV600E mutated DNA in plasma of melanoma patients. Clin Chim Acta 411:1319–1324

26. Davies H, Bignell GR, Cox C et al (2002) Mutations of the BRAF gene in human cancer. Nature 417:949–954

27. Vdovichenko KK, Markova SI, Belokhvostov AS (2004) Mutant form of BRAF gene in blood plasma of cancer patients. Ann N Y Acad Sci 1022:228–231

28. Cradic KW, Milosevic D, Rosenberg AM et al (2009) Mutant BRAF(T1799A) can be detected in the blood of papillary thyroid carcinoma patients and correlates with disease status. J Clin Endocrinol Metab 94:5001–5009

29. Chuang TC, Chuang AY, Poeta L et al (2010) Detectable BRAF mutation in serum DNA samples from patients with papillary thyroid carcinomas. Head Neck 32:229–234

Chapter 14

Gene Expression Analysis by qPCR in Clinical Kidney Transplantation

Michael Eikmans, Jacqueline D.H. Anholts, and Frans H.J. Claas

Abstract

Patients with a kidney transplant may encounter chronic dysfunction of their graft. Once damage in the graft has established, therapeutic intervention is less efficient. Clinical parameters and morphologic evaluation of biopsies are used for determining diagnosis and prognosis of the patient. Quantitative polymerase chain reaction (qPCR) may be integrated in clinical practice to facilitate routine diagnostics, risk assessment with respect to graft outcome, and determination of the response to therapy by the patient. The success of qPCR assays is highly dependent on the adequacy of the methodological procedures performed. Here, we describe tips and tricks for processing patient material, RNA analysis, and qPCR primer design and gene expression analyses.

Key words mRNA, Transplant, Kidney, Diagnosis, Prognosis

1 Introduction

The preferred treatment for patients suffering from end-stage renal disease is kidney transplantation. After transplantation, renal allograft rejection may affect graft outcome. At the long run, patients may suffer from chronic transplant dysfunction, which is accompanied by scar formation in the tissue and permanent loss of functional nephrons. This process is clinically characterized by progressive deterioration of renal function. With on-going chronic damage to the graft, therapeutic intervention for the patient is less efficient. It is therefore essential to predict clinical outcome at a time point before the development of overt scarring.

Clinical parameters and morphologic alterations in transplant biopsies are currently used in diagnostic practice and as means for determining prognosis of the patient. Improvement of diagnostic assessment and prediction of outcome in renal transplant patients may lead to earlier and more efficient strategies of therapeutic intervention. Messenger RNA measurement may be integrated in clinical practice to improve three fields of clinical transplantation practice:

Roberto Biassoni and Alessandro Raso (eds.), *Quantitative Real-Time PCR: Methods and Protocols*, Methods in Molecular Biology, vol. 1160, DOI 10.1007/978-1-4939-0733-5_14, © Springer Science+Business Media New York 2014

routine diagnostics, assessment of prognosis, and determination of the response to therapy by the patient. Firstly, molecular analysis of the tissue by mRNA expression analysis may help improve diagnostic accuracy [1–3]. Secondly, mRNA assessment could be used as prognostic tool: transcript levels may serve as a complement to histological findings for assessing risk of graft loss [4–7]. Thirdly, analysis of gene expression profiles may represent a means to predict the response of the patient to therapy: we [8, 9] and others [5, 10] have found particular expression profiles during acute transplant rejection, which are associated with therapy resistance of the patient. Fourthly, mRNA assessment may be used as a tool to monitor the extent of therapy-related negative side effects over time [11, 12]. Fifthly, gene expression profiles may enhance identification of patients who are eligible for weaning of immunosuppressive medication and who are prone to develop immunologic tolerance toward their graft [13–17].

To reach the goals outlined above, biomarkers are being established in renal transplant biopsies, urine, and peripheral blood to detect acute rejection and to provide information regarding risk of graft loss [18]. Microarray technology gives the possibility of simultaneously analysis of the RNA expression of all known genes in the genome in an unbiased manner. For analysis of a more limited set of genes of interest in a higher number of patient samples, application of quantitative polymerase chain reaction (qPCR) is more useful. This can be carried out with a specific probe sequence containing a reporter and quencher dye. Alternatively, compounds such as SYBR Green can be used in the reaction mixture, which emits light only in conditions where it is bound to double-stranded (amplified) nucleic acids.

Messenger RNA assessment in clinical biopsies for the applications mentioned above requires optimal protocols for tissue processing, RNA analysis, and qPCR assays. The success and the amplification signals of the PCR assays are highly dependent on the quality of the RNA, and on the efficiency of the RNA extraction and cDNA synthesis procedures, which precede the PCR. In the next sections we will describe methodology for the steps involved in clinical application of gene expression analysis: from processing of the patient material to the qPCR assays.

2   Materials

All equipment used in the steps before the actual qPCR assays, including thermal cyclers and pipettes, should be dedicated to pre-PCR work. This means that no DNA should be pipetted. Furthermore, to prevent aerosol formation and contamination as much as possible, we recommend using RNase-free, disposable, filtered pipette tips. Since human skin is a source of RNase molecules, gloves need to be worn while handling RNA, tubes, and pipettes.

2.1 Processing of Frozen Biopsy Material	1. Gloves. 2. Cryomicrotome for cutting sections. 3. Microscope glass slides. 4. Brush, for use in cryomicrotome. 5. Hematoxylin–eosin staining fluids. 6. Clean, RNase-free 2-ml microcentrifuge tubes (*see* **Note 1**). 7. Box of dry ice.
2.2 Processing of Blood Cells	1. Flow chamber. 2. Centrifuge. 3. Sterile 50-ml tubes. 4. Sterile 10-ml pipettes. 5. Plastic Pasteur pipettes (transfer pipettes). 6. Ficoll. 7. Sterile phosphate-buffered saline (PBS).
2.3 Processing of Urinary Sediment	1. Clean, RNase-free 1.5-ml microcentrifuge tubes (*see* **Note 1**). 2. Microcentrifuge. 3. Sterile PBS. 4. RNA*later* (Ambion, Invitrogen or Qiagen).
2.4 RNA Extraction	1. NucleoSpin miRNA kit (Macherey-Nagel) (*see* **Note 2**). This includes RNase-free recombinant DNase for on-column digestion of residual genomic DNA (*see* **Note 3**). 2. RNase-free 96–100 % ethanol. 3. Clean 1.5-ml microcentrifuge tubes (*see* **Note 1**). 4. Microcentrifuge. 5. Sterile, RNase-free water. 6. Thermal cycler.
2.5 Assessment of RNA Quantity and Quality	1. Automated electrophoresis system for RNA analysis (e.g., Experion from Bio-Rad). 2. Experion StdSens chips. 3. StdSens starter kit, including cleaning supplies, reagents, and RNA ladder. 4. Sterile water. 5. RNase-free 1.5-ml microcentrifuge tubes (*see* **Note 1**).
2.6 cDNA Synthesis	1. Reverse Transcriptase (*see* **Note 4**). 2. Oligo(dT)$_{15}$ primer and random nucleotide hexamers. 3. RNaseOUT recombinant ribonuclease inhibitor.

4. Dithiothreitol (DTT).

5. Reverse transcriptase buffer.

6. Deoxyribonucleotides triphosphate (dNTPs).

7. Thermal cycler.

8. Disposable plasticware (tubes, tips).

9. Micropipets.

**2.7 qPCR
Primer Design**

1. Computer with internet connection (*see* Subheading 3.7).

2.8 qPCR Analysis

1. Forward and reverse primers;

2. Intercalating supermix ready-to-use PCR buffer (Bio-Rad, Roche or Life Technologies); sterile water; cDNA.

3. Plastic tubes.

4. Single and multichannel repeater pipettes.

5. Optical 96-wells PCR plates.

6. Microseal film.

7. qPCR thermalcycler (Bio-Rad, Roche or Life Technologies).

---

3   Methods

Carry out all steps in a dedicated pre-PCR facility. The room should be devoid of PCR amplification products and DNA plasmids. Wear gloves at all times, especially while handling and opening tubes and reagents. Unless noted otherwise, the methods are performed at room temperature.

**3.1 Processing of
Frozen Biopsy Material**

1. Cut a 4-μm section from a snap frozen biopsy core using a cryomicrotome, which has been set to −20 °C.

2. Put the section on a glass slide, and stain with a fast hematoxylin–eosin (H and E) staining.

3. Check under the microscope whether the complete section contains renal cortical tissue.

4. Pre-cool clean, RNAse-free 2-ml-eppendorf tubes on dry ice.

5. Cut eight to ten 10-μm thick sections of renal cortex with a cryomicrotome from each frozen biopsy core (Fig. 1). If presence of renal medullary was suspected in the first tissue section (**step 3**), remove the medullary part from the thick sections (*see* **Note 5**).

6. Using a thin brush, gently put the thick sections into a 2-ml microcentrifuge tube (*see* **Note 6**).

7. Tubes with sectioned tissue need to be kept on dry ice without addition of RNA-preserving compounds. Add RNA lysis buffer within one hour after sectioning (*see* **Note 7**).

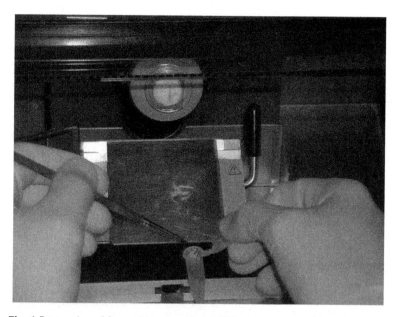

Fig. 1 Processing of frozen biopsy material. Thick tissue sections are cut with a cryomicrotome. The tissue is then transferred to a tube for RNA extraction and qPCR purposes

3.2 Processing of Blood Cells

The experiments need to be performed in a flow chamber. The methodology described below is suitable for obtaining lymphocytes from blood samples.

1. Use 10–80 ml sodium- or heparin blood.

2. Transfer a maximum of 20 ml blood to a siliconized 50-ml tube.

3. Dilute the blood 1:1, by adding an equal volume of PBS.

4. Gently put 13 ml Ficoll underneath the blood. Remove the pipette without dripping any fluid.

5. Centrifuge for 15 min at 800–1,200 ×g, with the break off (*see* **Note 8**).

6. Accurately transfer the Ficoll-ring (containing lymphocytes) to a new 50-ml tube, using a Pasteur pipette. Be sure not to transfer any Ficoll.

7. Fill up to 50 ml with PBS.

8. Centrifuge for 10 min at 800–1,200 ×g, with the break off.

9. Pour off the supernatant.

10. Resuspend all cells from one sample together in one vial with 11 ml PBS (*see* **Note 9**).

11. Fill up to 50 ml with PBS.

12. Centrifuge for 10 min at 800–1,200 ×g, using light breaking at the end.

13. Pour off the supernatant.

14. Add 50 μl RNA*later* to a maximum of 10×10^6 cells. Do not pipette up and down the pellet (*see* **Notes 10** and **11**).

3.3 Processing of Urinary Sediment

1. Centrifuge urine for 10 min at $2,800 \times g$.

2. Resuspend the pellet in 0.9 ml PBS by pipetting up and down for 3–4 times. Transfer the content to a 1.5-ml capped-tube.

3. Spin down pellet in microcentrifuge for 2 min at full speed.

4. Remove PBS supernatant as carefully as possible, without dispersing the pellet.

5. Add 25 μl RNA*later* to the cell pellet. Do not pipette up and down the pellet. Store vial at −20 °C (*see* **Note 10**).

3.4 RNA Extraction (See Notes 2 and 12)

1. Add 300 μl Buffer ML to a maximum of 30 mg tissue or 10^7 cells.

2. Vortex and incubate for 5 min at room temperature.

3. Place a NucleoSpin filter (violet ring) into a 2-ml collection tube.

4. Centrifuge for 1 min at $11,000 \times g$ in a microcentrifuge. This will help clearing the lysate from undissolved debris. Discard the NucleoSpin filter.

5. Add 150 μl 96–100 % ethanol to the lysate. Vortex immediately for 5 s, than incubate for 5 min at room temperature.

6. Place a NucleoSpin RNA column (blue ring) in a 2-ml collection tube. Load the column. Centrifuge for 1 min at $14,000 \times g$.

7. Save the flow-through (this contains the small RNAs).

8. Transfer the NucleoSpin RNA column into a new 2-ml collection tube.

9. Add 350 μl Buffer MDB to the RNA column. Centrifuge for 1 min at $11,000 \times g$.

10. Add 100 μl rDNase onto the membrane of the RNA column (*see* **Note 3**). Incubate for 15 min at room temperature, while leaving the lid open.

11. In the meantime add 300 μl Buffer MP to the saved flow-through from **step 7**.

12. Vortex for 5 s. Centrifuge for 3 min at $11,000 \times g$. This pellets the protein content.

13. Place a NucleoSpin Protein Removal column (white ring) in a 2-ml collection tube.

14. Pour the supernatant onto the column. Centrifuge for 1 min at $11,000 \times g$.

15. Add 800 μl Buffer MX. Vortex for 5 s.

16. Load 600 µl sample onto the column, containing the large RNA from **step 10**. Centrifuge 30 s at 11,000×*g*. Discard flow-through.

17. Repeat **step 16** two times to load the remaining sample.

18. Add 600 µl Buffer MW1 to the column. Centrifuge for 30 s at 11,000×*g*. Discard flow-through.

19. Add 700 µl Buffer MW2 to the column. Centrifuge for 30 s at 11,000×*g*. Discard flow-through.

20. Add 250 µl Buffer MW2 to the column. To dry the membrane completely, centrifuge for 2 min at 11,000×*g*. Discard flow-through.

21. Place the column into a new 1.5-ml collection tube.

22. Add 50–100 µl RNase-free water to the column (*see* **Note 13**). Incubate for 5 min at room temperature. Centrifuge for 30 s at 11,000×*g*. Store RNA at –80 °C or proceed immediately to the next sections.

3.5 Assessment of RNA Quantity and Quality

With respect to determining RNA quantities, automated electrophoresis (described below) gives comparable results to a NanoDrop, especially in the 20–100 ng/µl range (Fig. 2a). Generally, a reliable RNA quality value can be obtained with >20 ng/µl of RNA as input. With lower amounts of RNA it is advisable to use HighSens chips.

1. To clean the electrodes of the Experion, place a cleaning chip (contains 800 µl electrode cleaner) in the electrophoresis station. Close the lid for 2 min. Remove the chip.

2. Place another cleaning chip (containing 800 µl water) into the station. Close the lid for 5 min. Repeat this step. Leave the lid open for 1 min to dry.

3. Centrifuge 600 µl RNA gel (green cap) in a spin filter for 10 min at 1,500×*g*.

4. Pipette 65 µl filtered gel into an RNase-free tube. Add 1 µl RNA stain (blue cap). Vortex.

5. Pipet in separate tubes: at least 2 µl RNA ladder and 2 µl RNA sample. Incubate samples for 2 min at 70 °C. Immediately place on ice for 5 min.

6. Place an RNA StdSens chip on the platform. Add 9 µl gel-stain solution to the gel priming well.

7. Close the lid. Set pressure to "B" and time to "1." Press "start" button.

8. Check the backside of the chip for air bubbles.

9. Pipet 9 µl gel-stain solution into the well labelled "GS."

10. Pipet 9 µl filtered gel into the well labelled "G."

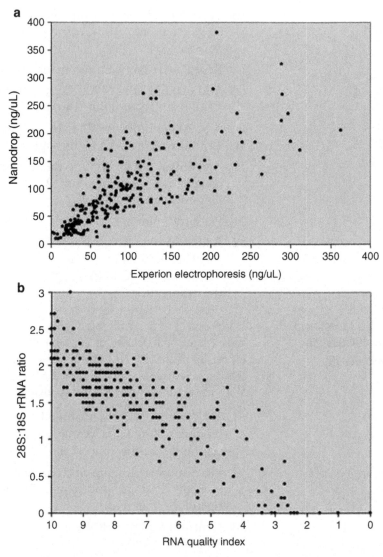

Fig. 2 Assessment of RNA quantity and quality. Results are shown for RNA that was obtained from different clinical samples (biopsies, blood cells). (**a**) A fairly high correlation is seen between RNA quantities measured by a NanoDrop and RNA quantities measured by gel electrophoresis (Experion), especially in the 20–100 ng/μl range. (**b**) The RNA quality index and the ratio of the 28S- to 18S ribosomal RNA product correlate with each other. Both parameters were assessed by the Experion system

11. Pipet 5 μl loading buffer (yellow cap) into each of the 12 remaining wells and into the well labelled "L" (ladder well).

12. Pipet 1 μl denatured RNA ladder into the ladder well (*see* **Note 14**).

13. Pipet 1 μl denatured sample into each of the 12 sample wells (*see* **Note 14**). Pipet 1 μl water into any unused sample wells.

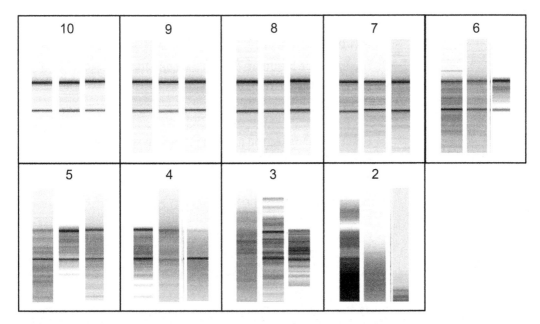

Fig. 3 RNA quality index as reflection of integrity of the RNA. The pictures show results for electrophoresis of RNA that was obtained from different clinical samples (biopsies, blood cells). The upper and lower band represent the 28S rRNA and 18S rRNA products, respectively. Examples are shown for various RNA samples with integrity matching different round numbers (from 10 down to 2) on the quality scale

14. Vortex the chip for 1 min.

15. Immediately run the chip in the electrophoresis station.

16. Each sample should show the 28S- and 18S ribosomal RNA products on the gel. In high-quality RNA the 28S–18S ratio is around 2.2. This ratio decreases with higher extent of RNA degradation. Although the integrity of the RNA is best reflected by the quality index, which runs from 10 (optimal) to 1 (highly degraded), we have seen a fairly high correlation between this index and the 28S–18S ratio (Fig. 2b). Generally, lower index values correspond with a higher extent of smear in the region between the 28S and 18S rRNA bands (Fig. 3).

3.6 cDNA Synthesis

1. Add together per reaction: 11 μl RNA/H$_2$O (maximum of 1 μg; *see* **Note 15**), 0.5 μl oligodT (0.5 μg/μl), 0.5 μl random nucleotide hexamers (0.5 μg/μl), and 1 μl 10 mM dNTP (*see* **Note 16**).

2. Incubate for 5 min at 65 °C.

3. In the meantime, prepare per reaction a mixture containing: 1 μl RNAseOUT rRNAse inhibitor (40 U/μl), 1 μl 0.1 M DTT, 1 μl SuperScript III (200 U/μl), and 4 μl 5× reverse transcriptase buffer.

4. Put the first mixture (containing the RNA), from **step 1**, on ice.

5. Add the two mixtures together. The total volume is now 20 μl.

6. Incubate for 5 min at 25 °C.

7. Incubate for 1 h at 50 °C.

8. Incubate for 5 min at 70 °C to terminate the reaction.

9. Dilute and store the cDNA samples (*see* **Note 17**).

3.7 qPCR Primer Design

1. Go to http://www.ensembl.org.

2. Enter the name of the gene of interest in the human database.

3. Select "Gene" under feature type, and select in the next field "Gene ID."

4. Selecting the right protein coding sequence in the transcript table enables to click on "Exons."

5. Sequence of the individual exons can be copy-pasted one by one to a Word document. Black-colored sequences represent exons, blue-colored sequences introns, and purple-colored untranslated regions.

6. Make a note in the mRNA sequence of where the introns would be located, and also of the length of the introns.

7. Open Primer3 (v. 0.4.0) at http://frodo.wi.mit.edu.

8. Copy-paste the complete exon sequence in the text box. For real-time PCR purposes, we set "product size ranges" to "80–200," and "number to return" to "50."

9. Try to find two to three different primer pairs, preferably targeting different regions in the cDNA, except the untranslated regions. When RNA is not treated with DNase (*see* Subheading 3.4), make sure that the forward and reverse primers span an intron of at least 800 base pairs (*see* **Note 18**).

10. Check primer pairs for cross-hybridization with other genes using NCBI/Primer-BLAST at http://www.ncbi.nlm.nih.gov/tools/primer-blast. If there is cross-hybridization, go back to **step 8**. A total of 10 mismatches or more in the primer set is acceptable.

11. Order at least two different primer pairs (*see* **Note 19**). Test specificity of the primers sets (*see* **Note 20**) by qPCR and melting curve analysis (*see* Subheading 3.8 for protocols), using cDNA samples and genomic DNA samples (*see* **Note 21**). Examples of optimal and nonoptimal melting curves (indicating low specificity) are shown in Fig. 4.

3.8 qPCR Analysis

1. Mix per reaction: 0.6 μl primers (mixture of both forward and reverse, each 25 μM), 7.5 μl ready-to-use PCR buffer (*see* **Note 22**), and 3.9 μl water, to reach a volume of 12 μl.

2. Pipet the reaction mixture on a PCR plate, using a single-channel repeater pipette.

3. Add 3 μl 1:25 diluted cDNA (*see* **Note 17**) per reaction.

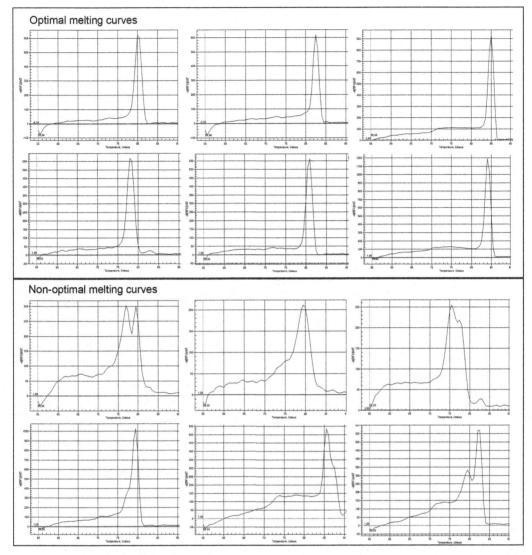

Fig. 4 PCR primer specificity as indicated by melting curve analysis. The figure shows eight examples of an optimal melting curve, indicating high specificity of the primers (*upper panel*). The *lower panel* shows eight examples of a nonoptimal melting curve, indicating moderate to low specificity of the primers

4. Cover the PCR plate with optical seal.

5. Incubate the plate in a thermal cycler at the following temperatures: 10 min at 95 °C; 40 cycles of 15 s at 95 °C and 1 min at 60 °C. At the end, a melting analysis needs to be performed (*see also* Subheading 3.7 and Fig. 4). This is performed by increasing the temperature from 55 to 95 °C with steps of 0.5 °C increment each for 10 s.

6. As standardization of the level of expression, reference genes need to be taken along for PCR (*see* **Note 23**).

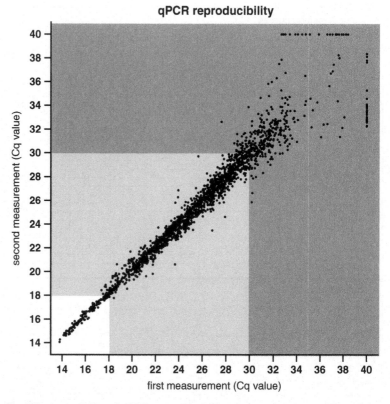

Fig. 5 Reproducibility of qPCR assessment. *Cq* values were obtained from 2,000 different duplicate measurements. Variation is smallest in samples with a *Cq* of 18 or lower (*white* area). Variation is increased in samples with a *Cq* between 18 and 30 (*light-grey* area), and is largest in samples with a *Cq* of 30 or higher (*dark-grey* area)

7. It is advisable to run at least a duplicate for each sample. In case the duplicates differ more than 0.5 Cq values from each other, a third run may be performed for that particular sample, and in the end the measurement differing the most from the other two should be discarded. We have provided information in Fig. 5 on the reproducibility of qPCR assessment: variation between duplicate qPCR measurements generally increases when the Cq value of the sample is higher.

4   Notes

1. It is best to use a dedicated bag of microcentrifuge tubes for RNA work, which means that gloves need to be worn at all time while reaching into the bag. It is not necessary to autoclave the tubes. Make sure that the tubes by the supplier are guaranteed RNase-free and sterile.

2. In the past we used RNeasy columns from Qiagen. Over the last years, microRNAs have gained attention because of their role in disease. Since small RNAs (<200 nucleotides) are for a large part lost with the RNeasy columns [19], we now are using a kit that is suitable for isolating both small and large RNAs. The yields of conventional (large) mRNA obtained with this kit are equal to those obtained with RNeasy [19].

3. Most of the times DNase treatment is performed on the RNA extraction columns. It is advisable not to treat the RNA with DNase treatment in case of extraction from urinary sediments, since in that case it may negatively affect PCR results. In that case, amplification of genomic DNA traces needs to be avoided by selecting primer pairs that span large intronic sequences (*see* Subheading 3.7).

4. Use Reverse Transcriptase that provides high yields of full-length cDNA product in order to achieve a more complete gene product representation (Superscript III or similar product).

5. Since biopsies are generally performed under guidance of ultrasound, most biopsy cores will consist primarily of cortical tissue. If one is interested in transcriptional regulation in the renal cortex only, it is essential to remove any renal medullary tissue as adequate as possible. We have noticed that expression levels of many transcripts differ considerably between the two compartments.

6. Try to work swiftly at this step, to keep the tube as cold as possible while putting the tissue sections in. Also, after removing it from the dry ice, hold on to the opened lid of the tube to prevent warming-up of the vessel wall by the body temperature of the hands.

7. Even with biopsy cores that had been fixed in optimal cutting temperature (OCT) compound on cork for up to 15 years we most of the times have been able to obtain high-quality RNA (RNA quality index ≥8) from the sections [8]. For best quality of RNA add lysis buffer shortly after sectioning. It is essential that the sectioned tissue is not thawed before adding the lysis buffer from the RNA extraction procedure. Thawing of frozen tissue in the absence of RNA-preserving buffer or RNA lysis buffer leads to release of intracellular RNase molecules and significant loss of RNA quality.

8. Do not use the break at the end of centrifuging in order to maintain the lymphocyte ring intact.

9. At this point, a small sample can be set apart to count the number of cells, using Türk solution and a Bürker counting chamber.

10. RNA*later* needs to cover the pellet completely. With bigger pellets, the RNA*later* volume added can be doubled. The original protocol of the supplier prescribes that, after adding the

RNA*later*, the tube is first incubated at 4 °C for 24 h before storing it at lower temperatures. We have found that this is not necessary, as long as the samples are subsequently stored at −20 °C or lower (e.g., −80 °C) for at least 24 h. Once in RNA*later*, the pellet can be thawed and frozen without interfering with the quality of the intracellular RNA.

11. Alternatively, the cells can be stored in RPMI culture medium and DMSO in liquid nitrogen, using conventional freezing procedures. In this way, RNA is equally well preserved as in RNA*later* [19]. RNA can still be extracted from the cells at a later time point, but then the cells need to be thawed first and washed, to get rid of the DMSO. This is necessary, since DMSO disturbs efficient RNA extraction.

12. Do not remove any of the RNA*later* from the sample before adding the lysis buffer (from the RNA extraction kit). Doing this may negatively affect RNA integrity. For the extraction of RNA from the sources of material described in this paper, sections do not need to be homogenized by a tissue lyser after addition of the lysis buffer. A few times of vortexing of the tube and incubation for 5 min at room temperature are sufficient. Using the NucleoSpin columns, we normally isolate small and large RNA in one fraction (option "L+S" in the kit).

13. If low RNA yields are expected, elution volume may be decreased to 30 μl.

14. Make sure that the amount loaded into the well does not exceed 1 μl. For best results, hold the pipette in a 30° angle while loading, and do not touch the bottom of the well with the pipette tip. To prevent air bubbles, do not pipette out completely, but only until the first stop.

15. To safe RNA, adding a maximum of 0.5 μg RNA to the reaction is sufficient. While cDNA reactions generally can handle a maximum of 1–1.5 μg of total RNA as input, addition of 1 μg RNA does not lead to a proportional increase in the amount of signal obtained by PCR on the cDNA. With an expected RNA yield <15 ng/μl, add 11 μl of undiluted RNA to the reaction.

16. When performing cDNA synthesis for a lot of samples at the same time, it is more convenient to carry out the reactions in PCR plates. For this, reaction mixture needs to be made for the number of samples intended (plus 10 % extra), and then this mixture is dispersed over the plate with a single-channel repeater pipette. After adding RNA to the individual wells, the plate is incubated in a thermal cycler for synthesis of the cDNA. The thermal cycler is dedicated to pre-PCR work and has pre-set temperature settings.

17. Repeated freezing–thawing cycles of stored cDNA samples leads to a decrease in PCR signal. We usually take out half of

the cDNA volume to make a 1:25 working solution (with water), and store the other half at –20 °C. Especially when running multiple PCR runs in a relatively short time period (days to weeks), it is best to keep the cDNA working solution at 4 °C. Make sure to properly cap the wells, so that fluid does not evaporate during storage.

18. Alternatively, one of the primers may be selected in such a way that one third to half of its 3′-end falls over an exon–intron boundary. For preventing genomic DNA amplification in the PCR, we however have had more success selecting each primer on a separate exon.

19. After having verified that primers do not cross-hybridize with other gene products, primer sequences can be obtained from a supplier (e.g., Sigma or Eurogentec). We have often noticed in PCR considerable differences in performance (specificity; signal intensity according to Cq values) between different primer pairs, which beforehand in theory should not have differed. It is therefore advisable to order and test multiple primer pairs per gene, and in the end pick the best one for application on patient samples.

20. To check for specificity of the primers, a melting curve analysis needs to be performed when using SYBR Green (*see also* Subheading 3.8). High specificity is indicated by the presence of one single, sharp peak (Fig. 4). Although the melting curve analysis is a good indicator of primer specificity, it is advisable to put the PCR product on an agarose gel: a single product of the right size should then be seen.

   Alternatively, real-time PCR reactions can be run using a fluorescently labelled probe, which binds to the template sequence located in between the forward and reverse primer. Running reactions with a probe enhances specificity of the assay, but also increases costs.

21. Make sure to include cDNA samples that have relatively high expression of the transcript of interest. Our interest lies mainly in immunologic markers, so we normally use cDNA derived from spleen, tonsil, and lymph nodes. If high expression of the particular transcript is expected in parenchymal cells of peripheral organs (liver, kidney etc), human reference total RNA (from Clontech; France or Cell Applications, Inc; San Diego, CA) provides an ideal source of starting material. PCRs on RNA, which is obtained commercially or derived from cell cultures, often give "cleaner" results than RNA from patient material. Therefore, during the test phase of the primers it is best to include also cDNA from clinical samples: we have occasionally observed that primers gave a single, sharp melting peak on commercial cDNA, whereas the same primers gave suboptimal results on clinical samples. We usually obtain

genomic DNA by Qiagen elution columns, and use 10 ng per reaction. In the PCR, the primers should give a Cq value on the DNA higher than 30, preferably higher than 35.

22. Many companies supply ready-to-use PCR buffer, which contains Taq polymerase enzyme, magnesium, and SYBR Green. If magnesium already is present in the buffer, be sure to check the end concentration in the reaction; generally this will be 3 mM. Some reactions may require higher concentrations, and therefore extra magnesium needs to be added.

23. There has been a lot of debate concerning the right choice of reference genes for mRNA analysis. In gene expression studies of cultured cells we use *GAPDH* and *β-actin*, and calculate the geometric mean of their levels. For mRNA studies in patient material we advise to include at least three reference genes. In a study on human biopsies we analyzed 18S rRNA, *GAPDH*, *β-actin*, and *HPRT-1*. The latter was eventually left out, since its levels showed the lowest correlation with the other three [8].

Acknowledgement

Part of the results shown in Fig. 5 were derived from a project on B cells in kidney transplantation, initiated by Dr. Sebastiaan Heidt (Leiden University Medical Center, Department of Immunohematology).

References

1. Donauer J, Rumberger B, Klein M et al (2003) Expression profiling on chronically rejected transplant kidneys. Transplantation 76:539–547

2. Koop K, Bakker RC, Eikmans M et al (2004) Differentiation between chronic rejection and chronic cyclosporine toxicity by analysis of renal cortical mRNA. Kidney Int 66:2038–2046

3. Reeve J, Einecke G, Mengel M et al (2009) Diagnosing rejection in renal transplants: a comparison of molecular- and histopathology-based approaches. Am J Transplant 9:1802–1810

4. Eikmans M, Sijpkens YW, Baelde HJ et al (2002) High transforming growth factor-beta and extracellular matrix mRNA response in renal allografts during early acute rejection is associated with absence of chronic rejection. Transplantation 73:573–579

5. Sarwal M, Chua MS, Kambham N et al (2003) Molecular heterogeneity in acute renal allograft rejection identified by DNA microarray profiling. N Engl J Med 349:125–138

6. Mengel M, Reeve J, Bunnag S et al (2009) Molecular correlates of scarring in kidney transplants: the emergence of mast cell transcripts. Am J Transplant 9:169–178

7. Sis B, Jhangri GS, Bunnag S et al (2009) Endothelial gene expression in kidney transplants with alloantibody indicates antibody-mediated damage despite lack of C4d staining. Am J Transplant 9:2312–2323

8. Rekers NV, Bajema IM, Mallat MJ et al (2012) Quantitative polymerase chain reaction profiling of immunomarkers in rejecting kidney allografts for predicting response to steroid treatment. Transplantation 94:596–602

9. Rekers NV, Bajema IM, Mallat MJ et al (2013) Increased metallothionein expression reflects steroid resistance in renal allograft recipients. Am J Transplant 13(8): 2106–2118

10. Desvaux D, Schwarzinger M, Pastural M et al (2004) Molecular diagnosis of renal-allograft rejection: correlation with histopathologic

evaluation and antirejection-therapy resistance. Transplantation 78:647–653

11. Flechner SM, Kurian SM, Solez K et al (2004) De novo kidney transplantation without use of calcineurin inhibitors preserves renal structure and function at two years. Am J Transplant 4: 1776–1785

12. Roos-Van Groningen MC, Scholten EM, Lelieveld PM et al (2006) Molecular comparison of calcineurin inhibitor-induced fibrogenic responses in protocol renal transplant biopsies. J Am Soc Nephrol 17:881–888

13. Brouard S, Mansfield E, Braud C et al (2007) Identification of a peripheral blood transcriptional biomarker panel associated with operational renal allograft tolerance. Proc Natl Acad Sci U S A 104:15448–15453

14. Martinez-Llordella M, Lozano JJ, Puig-Pey I et al (2008) Using transcriptional profiling to develop a diagnostic test of operational tolerance in liver transplant recipients. J Clin Invest 118:2845–2857

15. Newell KA, Asare A, Kirk AD et al (2010) Identification of a B cell signature associated with renal transplant tolerance in humans. J Clin Invest 120:1836–1847

16. Sagoo P, Perucha E, Sawitzki B et al (2010) Development of a cross-platform biomarker signature to detect renal transplant tolerance in humans. J Clin Invest 120:1848–1861

17. Bohne F, Martinez-Llordella M, Lozano JJ et al (2012) Intra-graft expression of genes involved in iron homeostasis predicts the development of operational tolerance in human liver transplantation. J Clin Invest 122:368–382

18. Eikmans M, Roelen DL, Claas FH (2008) Molecular monitoring for rejection and graft outcome in kidney transplantation. Expert Opin Med Diagn 2:1365–1379

19. Eikmans M, Rekers NV, Anholts JD et al (2013) Blood cell mRNAs and microRNAs: optimized protocols for extraction and preservation. Blood 121:e81–e89

Chapter 15

Posttranscriptional Regulatory Networks: From Expression Profiling to Integrative Analysis of mRNA and MicroRNA Data

Swanhild U. Meyer, Katharina Stoecker, Steffen Sass, Fabian J. Theis, and Michael W. Pfaffl

Abstract

Protein coding RNAs are posttranscriptionally regulated by microRNAs, a class of small noncoding RNAs. Insights in messenger RNA (mRNA) and microRNA (miRNA) regulatory interactions facilitate the understanding of fine-tuning of gene expression and might allow better estimation of protein synthesis. However, in silico predictions of mRNA–microRNA interactions do not take into account the specific transcriptomic status of the biological system and are biased by false positives. One possible solution to predict rather reliable mRNA-miRNA relations in the specific biological context is to integrate real mRNA and miRNA transcriptomic data as well as in silico target predictions. This chapter addresses the workflow and methods one can apply for expression profiling and the integrative analysis of mRNA and miRNA data, as well as how to analyze and interpret results, and how to build up models of posttranscriptional regulatory networks.

Key words mRNA, miRNA, Microarray, Multiple linear-regression, TaLasso, Pathway analysis, Quantitative real-time PCR, Gene ontology, R language, Genomatix pathway system

1 Introduction

Gene expression is regulated at the posttranscriptional level by small noncoding RNA species. One prominent class of small noncoding RNAs are microRNAs (miRNAs), which are 19–24 nt in length [1, 2]. miRNAs are transcribed by RNA polymerase II from independent genes or represent introns of messenger RNA (mRNA) transcripts [3, 4]. In the canonical miRNA biogenesis, processing of primary miRNAs to precursor miRNAs (~70 nt) is catalyzed by DROSHA in complex with dsRNA-binding proteins [3, 5]. Alternatively, precursor miRNAs can be generated by splicing and debranching of introns (mirtrons) [4, 6]. Precursor miRNAs are exported to the cytoplasm and are further processed by DICER-TRBP complex to form the miRNA duplex (~20 base pairs).

Roberto Biassoni and Alessandro Raso (eds.), *Quantitative Real-Time PCR: Methods and Protocols*, Methods in Molecular Biology, vol. 1160, DOI 10.1007/978-1-4939-0733-5_15, © Springer Science+Business Media New York 2014

The miRNA strand, which is loaded into the miRNA-induced silencing complex leads to translational repression, destabilization, and degradation of target mRNAs [3, 7]. Target mRNAs are recognized by partial base pairing in the 3'-untranslated region [8, 9], within the protein coding sequence, or 5'-untranslated region [10, 11]. A predominant function of miRNAs is to negatively regulate gene expression by decreasing mRNA levels [12]. There is evidence of relatively few miRNAs having a positive effect on target gene expression in certain cellular conditions [13, 14].

Insights in mRNA and miRNA regulatory interactions facilitate the understanding of fine-tuning of gene expression and might allow better estimation of protein synthesis. As miRNA deregulation is a hallmark of several diseases [1] the understanding of miRNA–mRNA relations is of high interest from a scientific as well as medical [15] therapeutic point of view. Currently, computational prediction of miRNA targets is biased by a high false positive rate due to the short target site sequence. Moreover, certain mRNAs are not expressed or targeted in specific biological conditions. However, holistic experimental analyses of miRNA–mRNA relations are time consuming, as for example argonaute cross-linking immunoprecipitation [16]. Using mRNA as well as miRNA transcriptomics data together with results from target prediction algorithms can be a suitable first step in revealing mRNA-miRNA relationships based on real data. Several approaches have been suggested for the joint analysis of miRNA and mRNA data [17]. Integrating expression data increases the chance of identifying functionally relevant RNA-interactions. Thus, this chapter focuses on how to perform expression profiling and how to integrate mRNA and miRNA data together with in silico target predictions as well as how to interpret results of joint expression data analysis (Fig. 1).

2 Materials

2.1 Sample Preparation

1. *Cell culture*. Always use the appropriate base medium, e.g., Dulbecco's Modified Eagle's Medium, and the corresponding supplements, e.g., amino acids and growth factors for culturing of cells. Growth conditions, such as 37 °C and 5 % CO_2 are suitable for most mammalian cells. A sterile working atmosphere and sterile working materials such as flasks, pipettes, filter tips etc. are mandatory.

2. *RNA extraction*. For total RNA purification you can utilize ready to use kits from Qiagen, Life Technologies, Promega, or PEQlab for example. We recommend the use of miRNeasy Mini Kit (Qiagen) as described by the supplier. Utilize RNase-free water for RNA elutions.

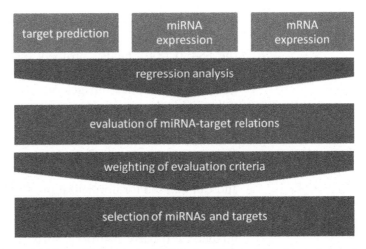

Fig. 1 Workflow of integrated mRNA and miRNA analysis. Integrated analysis starts with the input of target predictions and miRNA as well as mRNA data. Data from multiple regression analysis is further analyzed to make a final selection of miRNA–target interactions of high interest

3. *RNA quantity and quality.* RNA concentration and purity is determined by utilizing the Spectrophotometer NanoDrop1000 (Thermo Scientific) and RNA quality is analyzed using the Agilent 2100 Bioanalyzer (Agilent Technologies).

4. *cDNA synthesis and RT-qPCR.* Both, conversion of RNA to cDNA (miScript II RT Kit) and the quantification of microR-NAs (miScript SYBR Green PCR Kit and miScript Primer Assay) can be performed by utilizing the miScript System of Qiagen.

2.2 mRNA Profiling by Hybridization Arrays

For mRNA profiling an oligonucleotide hybridization-based plat-form such as the Gene 1.0 ST Array System from Affymetrix can be used. The Affymetrix GeneChip Mouse Gene 1.0 ST Array detects 28 853 well annotated genes and consists of a square glass sub-strate enclosed in a plastic cartridge implying 770 317 distinct 25-mer oligonucleotide probes, based on the February 2006 mouse genome sequence (UCSC mm8, NCBI build 36) with an extensive acquisition of RefSeq, putative complete CDS GenBank transcripts, all Ensembl transcript classes and RefSeq NMs from human and rat [18].

Use Affymetrix GeneChip® Whole Transcript (WT) Sense Target Labeling Assay and the corresponding Kits: GeneChip® Eukaryotic Poly-A RNA Control Kit, GeneChip® WT cDNA Synthesis and Amplification Kit, GeneChip® Sample Cleanup Module, GeneChip® WT Terminal Labeling Kit, GeneChip Hybridization, Wash and Stain Kit, and GeneChip® IVT cRNA Cleanup Kit. With this Kit system, samples are labeled and hybridized by generating amplified and biotinylated sense-strand DNA targets from the whole expressed genome of interest. Utilize the Affymetrix GeneChip® Fluidics Station 450 for the 169 array format applying the FS450_0007 fluidics protocol of Affymetrix. Use Affymetrix GeneChip® Scanner 3000 7G or a higher version for scanning and generation of optical images of the hybridized chips which are called DAT files.

After image exposure proceed the generated CEL files by using the Affymetrix GeneChip® Operating System (GCOS), Affymetrix GeneChip® Command Console (AGCC). Affymetrix Expression ConsoleTM Software is part of the Affymetrix Power Tool (APT) and can be used for normalization. Perform statistical analysis and pathway analysis with GeneChip®-compatibleTM Software, NetAffx Analysis Center, Integrated Genome Browser (IGB) and further third-party platforms such as Multi Experiment Viewer (MeV) [19, 20] and R programming language.

2.3 microRNA Profiling by Hybridization Microarrays

miRNA profiling can be performed by using an oligonucleotide hybridization-based platform such as the Mouse miRNA Microarray Release 15.0, 8×15 K from Agilent Technologies. The array consists of one glass slide formatted with eight high-definition 15 K arrays containing probes for 696 distinct miRNAs based on Sanger miRBase (release 15.0) (Agilent Technologies product information). Samples are labeled and hybridized by using the miRNA Complete Labeling and Hybridization Kit (Agilent Technologies). Signal intensities are acquired by using the Agilent Microarray Scanner G2505C and further processed by applying the Feature Extraction Software 10.7.3.1 (Agilent Technologies). Agilent's Feature Extraction software automatically reads and processes raw microarray image files. The software finds and places microarray grids, rejects outlier pixels, accurately determines feature intensities and ratios, flags outlier pixels, and calculates statistical confidences (Agilent technologies product information). R programming language can be utilized for statistical analysis and visualization of the data.

2.4 microRNA Profiling by qPCR Low-Density Arrays

qPCR profiling is facilitated by the TaqMan Rodent MicroRNA Array cards A and B 144 (Applied Biosystems by Life Technologies) which comprehensively cover Sanger miRBase v10. Together both cards (A and B) contain a total of 585 miRNA assays. Moreover, each array contains six control assays (Applied Biosystems product information). TaqMan MicroRNA Arrays are used in conjunction with Megaplex™ RT Primers that are predefined pools of up to 381 RT primers. Megaplex™ PreAmp Primers are used for pre-amplification step. Low-density arrays (385-well format) are run on the 7900 HT Fast Real-Time PCR System (Applied Biosystems by Life Technologies). Quality control and derivation of Cq-values can be done using RQ Manager 1.2 (Applied Biosystems by Life Technologies). The data normalization, comprising analysis and visualization, can be performed by using RealTime StatMiner (Integromics) software as well as R programming.

2.5 Bioinformatics and Databases

For in silico target prediction use the latest version of TargetScan (currently TargetScan Release 6.2, June 2012) (http://www.targetscan.org/) [21, 22]. Moreover, use miRanda (currently the latest release of microrna.org is August 2010; http://www.microrna.org/) [23, 24] for target prediction.

For analyzing the inverse relation of expressed miRNA and mRNAs in conjunction with target predictions we recommend using a Lasso regression model [25]. miRNA–mRNA relations derived from the regression analysis can be further processed by testing for enrichment in gene ontology (GO) terms [26], or KEGG pathways (http://www.genome.jp/kegg/pathway.html) amongst others. Applying Genomatix Pathway System (GePS) (Genomatix) facilitates the comprehensive analysis and visualization of enriched canonical pathways, GO terms, disease terms, and transcription factors based on information extracted from public and proprietary databases [27] and co-citation in the literature. GePS also facilitates the creation and extension of networks based on literature data.

3 Methods

Sophisticated analysis of transcriptomics data together with bioinformatic data mining does not only reveal novel players in biological processes, but in addition generates a comprehensive view on posttranscriptional networks. Prior to generating transcriptomic regulatory networks including protein coding transcripts as well as small regulatory RNAs, the expression of mRNAs and microRNAs should be monitored. For expression studies it is important to use standardized conditions including reasonable numbers of technical and biological replicates and standard preparation protocols for sample processing as well as tangible aims of the study.

3.1 Sample Preparation

1. *Cell Culture*. Cell cultures are utilized in medical and molecular laboratories for diagnostics as well as research. In most cases, cells are cultivated for days or weeks to receive sufficient amounts of cells for analysis. For RNA extraction from cells the minimum amount of starting material is usually 100 cells [28]. The maximum amount of cells depends on the RNA content of the cell type. Parameters influencing the reproducibility of results are as follows:

   • Cell culture reagents and working materials.

   • Experience of the operator himself or herself.

For culturing of cells one should always use the same batch and number of cells and if possible even the same passage. It is also recommended to use cells not longer than 20 passages [29]. Some cells lose their characteristics rather rapidly when taken into culture. In these cases, only a few passages are advisable. In cell culture experiments technical replicates ($n > 2$) and biological replicates ($n > 3$) are usually utilized for the whole experiment. The vitality of cells should be monitored during the experiment. This can be visually verified by using microscopes and additionally with cell viability tests (e.g., Promega) or electric cell-substrate impedance sensing [30]. The reproducibility of results strongly depends on highly standardized workflows for each independent experiment. For RNA isolation you should also use equal volumes of cell lysis buffer. Lysed cells should always be kept on dry ice and stored at –80 °C.

2. *RNA Extraction.* Since both mRNA and miRNA should be included in expression profiling experiments, total RNA purification kits need to be used for isolating RNA from cultured cells or various tissues. It is very important that the kit recovers RNA molecules smaller than 200 nucleotides. Homogenized cell lysates should be thawed at 37 °C and subsequently incubated at room temperature (RT) for 5 min. The miRNeasy Mini kit (Qiagen) combines phenol/guanidine-based sample lysis with chloroform-based separation of nucleic acids from proteins and other cell constituents. For RNA purification silica membrane columns are used. Due to the toxicity of Phenol and Chloroform it is very important to work carefully under a flue. After sample homogenization with the Qiazol lysis reagent and the addition of chloroform the aqueous phase is separated from the organic phase by centrifugation. RNA is included in the upper aqueous phase while DNA is located in the interphase and proteins in the lower organic phase or interphase. Only the upper aqueous phase is separated and contamination with the interphase and lower phase should be strongly avoided. The RNA should be isolated as described by the supplier [28]. DNA digest on the spin column is also not recommended, because of material loss and less inefficiency compared to DNase treatment protocols of solved RNA. Total RNA is than eluted in 40 μl RNase-free water and stored at –80 °C. To obtain a higher total RNA concentration, it is recommended to repeat the elution step by using the same RNeasy spin column and the first eluate [28].

3. *RNA quantity and quality.* After RNA extraction, RNA concentration and purity are determined. Analyze 1.5 μl RNA solution by utilizing the Spectrophotometer NanoDrop1000. All RNA samples should occupy a 260/280 ratio within a range of 2.8–2.1. The RNA quality can be further confirmed

by using the Agilent 2100 Bioanalyzer. The Agilent 2100 Bioanalyzer is a chip-based platform that uses microcapillary electrophoresis to analyze proteins, nucleic acids or cells [31].

4. *cDNA synthesis and RT-qPCR.* Microarray technology (*see* Subheadings 3.2 and 3.3) is one of the most powerful tools for assumption based large-scale expression profiling. However, microarray profiling results have to be validated by using quantitative real-time PCR (qPCR). qPCR utilizes an RNA-dependent DNA polymerase to synthesize complementary DNA. For the reverse transcription of miRNA and mRNA we used a polyadenylation based approach (miScript II RT Kit by Qiagen). For the master mix add 4 µl 5× miScript HiFlex buffer, 2 µl 10× miScript Nucleic Mix, 2 µl miScript Reverse Transcriptase Mix, and RNA. The amount of RNA is in the range of 10 pg to 1 µg and depends on the experimental design and number of target reactions. Fill the reaction up with water to a total volume of 20 µl. The RT-reactions are incubated for 1 h at 37 °C and the cDNA reaction is stopped by incubation at 95 °C for 5 min. Samples should be stored at –20 °C [32].

The microRNA and gene expression can be analyzed with the CFX384 Touch 244 Real-Time Detection Cycler (Bio-Rad Laboratories). For specific quantification of miRNA and mRNA expression the miScript Primer Assay and the miScript SYBR Green-based RT-PCR Kit can be used. The reaction mix should be prepared as described by the supplier. The following cycling conditions should be used: after initial activation for 15 min at 95°, the cycle steps of denaturation for 15 s at 94 °C, annealing for 30 s at 55 °C, and an extension phase for 30 s at 70 °C should be repeated for 39 cycles [32] (*see* **Notes 1** and **2**).

3.2 mRNA Profiling and Validation

1. *mRNA hybridization microarrays.* Microarray technology uses the natural attraction between nucleotides, consisting of probe–target hybridization, detection and quantification of labeled targets to determine the relative amount of nucleic acid sequences and enables high-throughput analysis of transcripts. Therefore, the major applications of microarray technology are gene expression profiling and genetic variation analysis [33]. Using in situ synthesized oligonucleotides as probes and in silico designed microarrays, Affymetrix pioneered the understanding of total transcript activity. This chapter focuses on gene expression profiling with Affymetrix Gene Chip Gene 1.0 ST Array System that offers an option for whole-transcript coverage. When performing mRNA expression profiling by using Gene Chip Mouse Gene 1.0 ST Array (Affymetrix) follow the manufacturer's instructions [34] (*see* **Notes 3–5**). The manufacturer's instructions are summarized below.

2. *Isolation of total mature RNA from cells* (*see* Subheadings 2.1 and 3.1.)

3. *Preparation of total RNA with T7-(N)6 Primers and Poly-A RNA Controls.* For this step the GeneChip® Poly-A RNA Control Kit is required. In the first step the dilutions of Poly-A RNA Controls is prepared. Add 2 µl of 269 Poly-A RNA Control stock to 38 µl of Poly-A control Dil Buffer (first dilution 1:20). Mix gently and spin down the tube. Transfer 2 µl of the first dilution to 98 µl of Poly-A Control Dil Buffer (second dilution: 1:50). Transfer 2 µl of the second dilution to 98 µl of Poly-A Control Dil Buffer (third dilution 1:50). Next, prepare the solution of T7-(N)6 Primers/Poly-A RNA Controls. Add 2 µl of the T7-(N)6 Primers (stock conc. 2.5 µg/µl), 2 µl diluted Poly-A RNA Control (3rd dilution, 1:50) to 16 µl RNase-free water in a non-stick RNase-free tube. Mix the solution, spin down and place it on ice. Performing GenChip Gene 1.0 ST Arrays use 100 ng total sample RNA, add 2 µl of the prepared T7-(N)6 Primers/Poly-A RNA Controls solution and fill up to a total volume of 5 µl with RNase-free water. The mix is flicked, followed by centrifugation. The reaction mix is than incubated at 70 °C for 5 min and at 4 °C for 2 min. The mix is placed on ice.

4. *First-Cycle, First-Strand and second strand cDNA Synthesis.* The GeneChip® WT cDNA Synthesis Kit is required for this preparation step. For the first-cycle of first strand cDNA synthesis mix 2 µl of 5× 1st Strand Buffer, 1 µl DTT (0.1 M), 0.5 µl dNTP Mix (10 mM), 0.5 µl RNease Inhibitor, 1 µl SuperScript II enzyme in one tube. Add 5 µl of the master mix to the T7-(N)6 Primers/Poly-A RNA Controls solution, mix, and centrifuge the tube. The total volume for first strand cDNA synthesis is 10 µl. The reaction is incubated at 25 °C for 10 min, at 42 °C for 60 min, at 70 °C for 10 min, and finally at 4 °C for 2 min. Continue with the second strand cDNA synthesis and prepare the second strand master mix: To do so mix 4.8 µl RNase-free water with 4 µl $MgCl_2$ (17.5 mM), 0.4 µl dNTP Mix (10 mM), 0.6 µl DNA Polymerase I, and 0.2 µl RNAse H, and 10 µl of the first-cycle second-strand master mix was added to the reaction tube of the first-strand cDNA synthesis reaction. Gently vortex the tube and centrifuge. The reaction is incubated at 16 °C for 120 min, at 75 °C for 10 min, and at 4 °C for further 2 min.

5. *First-cycle, cRNA synthesis and cleanup.* This procedure requires the GeneChip® WT cDNA Amplification Kit and the GeneChip® Sample Cleanup Module. Prepare the IVT master mix, including 5 µl of 10× IVT, 20 µl IVT NTP Mix, and 5 µl IVT enzyme mix. Transfer this 30 µl of the IVT master mix to

the first-cycle cDNA synthesis reaction. Mix the sample again and centrifuge. The reaction is than incubated for 16 h at 37 °C. Proceed to the cleanup procedure for cRNA using the GeneChip Sample Cleanup Module. Add 50 µl of RNase-free water to each IVT reaction, followed by adding 350 µl of cRNA Binding Buffer to each reaction. Vortex the samples for 3 s. Add 250 µl of 100 % EtOH to each reaction and mix. Transfer the sample to the IVT cRNA Cleanup Spin Column and centrifuge for 15 s at $\geq 8,000 \times g$. Discard the flow-through. The spin column is transferred to a new 2 ml collection tube and the column is washed by adding 500 µl of cRNA wash Buffer. The column is centrifuged again for 15 s at $\geq 8,000 \times g$. Discard the flow-through. The column is washed again with 500 µl of 80 % EtOH. Leave the cap of the tube open and centrifuge the column for 5 min at $\leq 25,000 \times g$. The spin column is transferred to a new 1.5 ml collection tube and 15 µl of RNase-free water are added to the membrane of the column directly. Incubate the reaction for 5 min at room temperature. Subsequently, centrifuge for 1 min at $\leq 25,000 \times g$. The concentration can be determined by measuring the absorbance in a spectrophotometer (e.g., NanoDrop).

6. *Second-cycle, first-strand cDNA synthesis.* For this step the GeneChip® WT cDNA Synthesis Kit is used. In a strip tube the cRNA sample (max. 10 µg) is mixed with 1.5 µl Random Primers (3 µg/µl) and the reaction is filled up with RNease-free water to a total volume of 8 µl. Mix and spin down the tube. The cRNA/Random Primer mix is incubated at 70 °C for 5 min, followed by incubation at 25 °C for 5 min and at 4 °C for 2 min. In a second tube the second cycle, reverse transcription master mix is prepared. Combine 4 µl of 5× 1st Strand Buffer with 2.0 µl DTT (0.1 M), 1.25 µl dNTP+dUTP (10 mM), and 4.75 µl SuperScript II. This 12 µl of second-cycle, first-strand cDNA synthesis master mix is transferred to the second-cycle, cRNA/Random Primers Mix. After flicking, the tube is centrifuged briefly. The reaction is incubated at 25 °C for 10 min, at 42 °C for 90 min, at 70 °C for 10 min, and at 4 °C for 2 min.

7. *Hydrolysis of cRNA and cleanup of single-stranded DNA.* For this preparation step the GeneChip® WT cDNA Synthesis Kit and the GeneChip® Sample Cleanup Module are required again. 1 µl of RNase H is added to each sample and the reaction is incubated at 37 °C for 45 min, at 95 °C for 5 min, and at 4 °C for 2 min. Add 80 µl of RNase-free water to each sample, followed by adding 370 µl cDNA Binding Buffer. Vortex the sample for 3 s. Convey the total volume of 471 µl to a cDNA Spin Column from the GeneChip Sample Cleanup Module. Spin for 1 min at $\geq 8,000 \times g$. Discard the flow-through.

The spin column is transferred to a new 2 ml collection tube and 750 µl of cDNA Wash Buffer are added. Centrifuge the tube at $\geq$8,000×g for 1 min and discard flow-through. Open the cap of the cDNA Cleanup Spin Column and spin at $\leq$25,000×g for 5 min, discard the flow-through, and place the column in a new 1.5 ml collection tube. Add 15 µl of the cDNA Elution Buffer directly to the spin column membrane and centrifuge at $\leq$25,000×g for 1 min. Add another 15 µl of cDNA Elution Buffer to the column membrane and incubate at room temperature for 1 min, followed by spinning at $\leq$25,000×g for 1 min. Determine the yield by spectrophotometric UV measurement (e.g., NanoDrop).

8. *Fragmentation of single-stranded DNA.* This step requires the GeneChip® WT Terminal Labeling Kit. Use 0.2 ml strip tubes for the fragmentation. First, add 5.5 µg single-stranded DNA to a tube and fill up with RNase-free water to a total volume of 31.2 µl. Second, prepare the fragmentation master mix in a fresh tube, by adding 10 µl RNase-free water, 4.8 µl 10× cDNA Fragmentation Buffer, 1 µl UDG (10 U/µl), and 1 µl APE 1 (1,000 U/µl) per reaction. This 16.8 µl fragmentation master mix is added to the prior prepared single-stranded DNA, vortex gently and centrifuge the sample. The reaction is incubated at 37 °C for 60 min, at 93 °C for 2 min and in the end at 4 °C for 2 min. After incubation transfer 45 µl of the reaction mix to a new 0.2 ml strip tube. Examine the RNA quality using 2 µl of the residual RNA with the Bioanalyzer (as described in further detail in Chapter 5). The fragmented single stranded DNA should be approximately 40–70 nt.

9. *Labeling of fragmented single-stranded DNA.* The GeneChip® WT Terminal Labeling Kit will be used. First, prepare a labeling master mix by mixing 12 µl of the 5× TdT buffer, 2 µl TdT, and 1 µl DNA Labeling Reagent in a 0.2 ml tube and aliquot 15 µl of the prepared master mix (total volume 60 µl). After adding 15 µl of the master mix to the 45 µl fragmented single stranded DNA, the tubes are mixed and centrifuged. Incubate the reactions at 37 °C for 60 min, at 70 °C for 10 min, and finally at 4 °C for 2 min.

10. *Hybridization.* For hybridization the GeneChip Hybridization, Wash and Stain Kit is required. For the format array a total volume of 100 µl Hybridization Cocktail is needed. Add 27 µl fragmented and labeled DNA target, 1.7 µl (3 nM) Control Oligonucleotide B2, 5 µl of 20× Eukaryotic Hybridization Controls, 50 µl of 2× Hybridization Mix, 7 µl of DMSO, and 9.3 µl nuclease-free water to a 1.5 ml RNase-free tube. Gently vortex the tube and heat the hybridization cocktail at 99 °C for 5 min, subsequently cool the reaction for 5 min to 45 °C and centrifuge at full speed for 1 min. Equilibrate the GeneChip

ST Array to room temperature and inject 80 μl of the Hybridization Cocktail, including the fragmented and labeled DNA, into the array. The array is then placed into a rocker hybridization oven (45 °C, 600 rpm) for 17 h.

11. *Wash, Stain and Scan.* For washing, staining and scanning of the arrays, the Gene Chip Hybridization, Wash and Stain Kit is required. After hybridization vent the array and extract the Hybridization Cocktail. Refill the array with 100 μl wash buffer A. Gently tap the bottles of staining reagents and aliquot 600 μl of staining cocktail 1 into a 1.5 ml amber tube, 600 μl of staining cocktail 2 into a further (clear) microcentrifuge tube and 800 μl of Array Holding Buffer into a new 1.5 ml microcentrifuge tube. All tubes are centrifuged to remove any air bubbles. For washing and staining the probe array, the Fluidics Station 450/250 is used. Please, follow the detailed manufacturer's instructions [35].

12. *Scanning.* If there are no air bubbles inside the array, it is ready to be scanned on the GeneChip Scanner 3000. The scanner is also controlled by Affymetrix® GeneChip Command Console (AGCC). Optical images of the hybridized chip, called DAT files, are generated.

13. *Data file generation.* Proceed CEL files from the prior generated DAT files which contain intensity information by using the Affymetrix GeneChip® Operating System (GCOS) and Affymetrix GeneChip® Command Console (AGCC). The Affymetrix Expression ConsoleTM Software can be used for background adjustment, normalization, data quality control and gene-level signal detection, supporting also probe set summarization and CHP file generation for 3′ WT expression arrays.

14. *Statistical analysis and pathway analysis.* For the statistical as well as for the pathway analysis you can use 3rd party platforms as well as R programming language. The correlation coefficient can be calculated. High correlation coefficients >0.90, indicate a high reproducibility of the data. The analysis of variance (ANOVA) is extremely important during microarray processing. One-way analysis of variance tests significance of one condition, e.g., time, whereas two-way analysis of variance measures the significance of two factors simultaneously, e.g., time and treatment. For statistical analysis you can use different statistical test, e.g., the Mann–Whitney Test. It is also recommended to filter for low intensity values, because counts <30 are close to the background and therefore should be removed from the subsequent analysis. For visual illustration of the data you can use heat maps and scatter plots. It is also recommended doing cluster analysis such as principle component analysis. Gene expression profiling for specific marker-genes should be

performed as well. For validation of microarray results it is also recommended to use quantitative real-time PCR (*see* **Notes 1, 2, 6–8**).

3.3 microRNA Profiling by Hybridization Microarrays

When performing miRNA expression profiling by using Mouse miRNA Microarrays (Agilent Technologies) (*see* **Note 9**) follow the manufacturer's instructions specified for the miRNA Complete Labeling and Hybridization Kit:

1. *Labeling.* Dephosphorylate and spike 100 ng total RNA per sample by mixing 4 μl of RNA with 0.7 μl 10× Calf Intestinal Phosphatase Buffer, 1.1 μl Labeling Spike-in, 0.5 nuclease-free water, and 0.7 μl Calf Intestinal Phosphatase. Incubate the reaction mixture at 37 °C for 30 min. Denature RNA by adding 5 μl DMSO at 100 °C for 8 min and cool at 0 °C for 5 min. For fluorophor ligation add 2 μl 10× T4 RNA Ligase Buffer, 2 μl nuclease-free water, 3 μl pCp-Cys, and 1 μl T4 RNA Ligase and incubate at 16 °C for 2 h. Purify samples by using Bio-Spin 6 Chromatography Columns (Bio-Rad) and dry in a SpeedVac Concentrator (Thermo Scientific) at 50 °C.

2. *Hybridization and scanning.* For hybridization of labeled miRNAs resuspend the dried samples in 17 μl nuclease-free water and mix with 1 μl Hybridization Spike-in, 4.5 μl 10× GE Blocking Agent, and 22.5 μl 2× Hi-RPM Hybridization Buffer and incubate at 100 °C for 5 min with subsequent cooling at 0 °C for 5 min and centrifugation. Load the samples into the array and hybridize at 55 °C for 20 h. Wash miRNA microarrays and scan in a single pass mode with a scan resolution of 3 μm, 20 bit mode. Extract signal intensities and background and log2 transform the data by using Feature Extraction Software 10.7.3.1 (Agilent Technologies).

3. *Evaluation and analysis of data.* Retain miRNAs that show a signal greater than zero in at least two of the replicates within one group. For miRNAs that meet these detection criteria define a value equal to zero as outlier and replace it by the mean value of the other replicates. Normalize the data by applying loess M normalization [36, 37]. Microarray data should be registered into ArrayExpress database [38], or Gene Expression Omnibus [39] which are publicly available repositories consistent with the MIAME guidelines [40]. For further processing of the data use GeneSpring GX Software (Agilent Technologies) or R programming (*see* **Note 10**). To detect the significant differential expression use significance analysis of microarray (SAM) [41]. SAM is an assumption free approach adapted to microarray analysis that identifies differentially expressed miRNAs by permutation and performing false discovery rate (FDR) correction of *p*-values.

**3.4   microRNA
Profiling by qPCR
Low-Density Arrays**

Validation of miRNA profiling results derived from hybridization-based arrays can be performed by qPCR based analysis (*see* **Note 11**). We recommend setting up three reverse transcription reactions per sample as the reverse transcription itself introduces bias. When using the TaqMan Rodent MicroRNA Array system follow the manufacturer's instructions.

1. *Reverse transcription.* Prepare the reverse transcription reaction mix by adding the following amounts per sample: 0.8 µl Megaplex RT Primers (10×), 0.2 µl dNTPs with dTTP (100 mM), 1.5 µl MultiScribe Reverse Transcriptase (50 U/µl), 0.8 µl RT Buffer (10×), 0.9 µl magnesium chloride (25 mM), 0.1 µl RNase Inhibitor (20 U/µl), 0.2 µl nuclease-free water. Add 1–350 ng total RNA or water for the no template control reaction, respectively. Incubate the reaction mixture on ice for 5 min. Using the 9700HT Systems (Applied Biosystems by Life Technology) set up the following run method: 40 cycles of 16 °C for 2 min, 42 °C for 1 min, and 50 °C for 1 s. Then set up a final step of 85 °C for 5 min followed by cooling to 4 °C.

2. *Pre-amplification.* Pre-amplify the reverse transcription product by mixing 12.5 µl TaqMan PreAmp Master Mix (2×), 2.5 µl Megaplex PreAmp Primers (10×), and 7.5 µl nuclease-free water and mix by inverting the tube. Subsequently add 2.5 µl reverse transcription product and 22.5 µl of the pre-amplification mixture and incubate on ice for 5 min. Set up the following running conditions on a 9700HT platform: 95 °C for 10 min, 55 °C for 2 min, 72 °C for 2 min followed by 12 cycles consisting of 95 °C for 15 s and 60 °C for 4 min. To facilitate enzyme inactivation heat it to 99.9 °C for 10 min and cool down to 4 °C. To dilute the pre-amplification product add 75 µl of 0.1× TE Buffer (8.0 pH) to each reaction. When having three reverse transcription reactions per sample we recommend to aequimolarly pool the reaction products by unifying 4 µl each (*see* **Note 12**).

3. *Real-time quantitative PCR.* Run the real-time PCR reaction by mixing the following volumes per array: 450 µl TaqMan Universal PCR Master Mix (no AmpErase UNG, 2×), 9 µl of the diluted and pooled pre-amplification product, 441 µl nuclease-free water. Load the array by dispensing 100 µl of the PCR reaction mix into each port of the TaqMan MicroRNA Array, centrifuge and seal the array. Run the array on the 7900HT System by choosing "Relative Quantification" and the 384-well TaqMan Low Density Array default thermal-cycling conditions of the SDS software.

4. *Evaluation and analysis of data.* To review the results, transfer the SDS files into a Comparative CT (RQ) study using the RQ Manager software (Applied Biosystems) (*see* **Note 13**). Applied Biosystems recommends analyzing the study with "Automatic

Baseline" and "Manual CT" set to 0.2. View each amplification plot manually and adjust the threshold settings for individual assays if necessary. It is important to use the same threshold settings across all samples or arrays within a study for a given assay. For detailed downstream analysis use software such as Real-Time StatMiner Software (Integromics) which amongst many other analysis options allows to for example omit miRNAs which do not show cycle of quantification (Cq) values smaller 32 in at least two of the corresponding replicates of a group. For normalization of the data use modified loess method [36] (*see* **Note 14**). Calculate fold changes of relative expressions (*see* **Note 8**) and identify significant differential expression by applying SAM and FDR correction.

3.5 Integrative Analysis of miRNA and mRNA Expression Data

The main assumption in the joint analysis of miRNA and mRNA expression data is that regulation of gene expression by miRNAs primarily takes place in the form of mRNA degradation. It was shown that this actually holds for mammalian cells [12]. Therefore, we can assume that changes in gene expression on mRNA level can be explained by expression changes of targeting miRNAs.

A straightforward approach for detecting these regulatory relations would be to calculate the Pearson correlation coefficient (r) between the mRNA expression and the expression of miRNAs, which are predicted to target the respective gene, over certain conditions. Afterwards one is able to determine the statistical significance of the correlation coefficient that can be utilized to select significantly anticorrelated miRNA–mRNA relations [42]. However, there is one major issue in the correlation analysis of mRNA and miRNA expression data, since a gene can also be targeted by more than one miRNA. A linear relationship between the expression profiles of several miRNAs and the expression of a common predicted target gene cannot be detected in a one-to-one fashion, because each of these miRNAs may have individual influences on the gene expression (Fig. 2). A more suitable method for quantifying the down-regulation by predicted miRNAs is therefore to use a multiple linear regression approach. Here, the goal is to simultaneously incorporate the multiple expression profiles of the predicted miRNAs in order to assess their individual contribution in explaining the mRNA expression. Since target predictions usually consist of many false positives, it is desirable to select for each gene only these miRNAs out of the set of predicted ones that have an actual influence on the mRNA expression. In order to achieve this a Lasso regression model can be used that performs a feature selection on the explanatory variables. The LASSO regression model is therefore suitable to identify potential regulatory relations between miRNAs and genes [25]. Since we are mainly interested in the down-regulation of genes, a Lasso regression model with non-positive constraints appears to be the most appropriate approach. This is implemented for example in TaLasso [17].

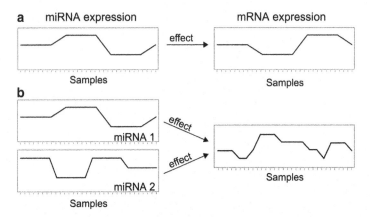

Fig. 2 MiRNA–target relations depending on target prediction algorithms. (**a**) The mRNA expression profile of a gene is expected to be anti-correlated to a single targeting miRNA over different conditions if the respective miRNA has a regulatory effect on the gene. (**b**) If more than one miRNA has a regulatory effect on the gene, the anticorrelation becomes less clear, since all targeting miRNAs can have individual influences on the mRNA expression

An overview of different approaches for the determination of miRNA–mRNA relations based on target predictions and combined expression data is given in the review of Muniategui et al. [17] (*see* **Notes 15** and **16**). The integrative analysis of mRNA and miRNA expression can be visualized by the open-source software Cytoscape 2.8.3 (http://www.cytoscape.org/) (Fig. 3).

TaLasso is available as an easy-to-use Web interface that can deal with miRNA and mRNA expression matrices. Two tab-separated text files must be provided as input. These files have to contain the respective expression values as well as the miRNA and mRNA identifier as row names. The columns of the expression matrices correspond to the samples and must be matched. Examples for such kind of input files are given on the Web site (http://talasso.cnb.csic.es/) (Fig. 4). Several target predictions resources can be chosen as underlying network. As mentioned above, we propose TargetScan as target predication tool (*see* **Note 17**). Furthermore, a set of experimentally validated target interactions can be selected in order to assess the amount of validated miRNA–mRNA relationships in the result. Standard settings can be used for all other parameters. After submitting the job, the result page will show up (Fig. 5). It contains all predicted target relationships that were predicted by the algorithm. The score and the *p*-value indicate the significance of the interaction. The result can be downloaded by right-clicking on the download links above the result list and selecting "Save as…". All three result files should be downloaded. The network of miRNA–mRNA interactions is stored as a comma-separated text file containing the assignment matrix. In order to prepare this matrix for importing into graph visualization tools, it has to be preprocessed.

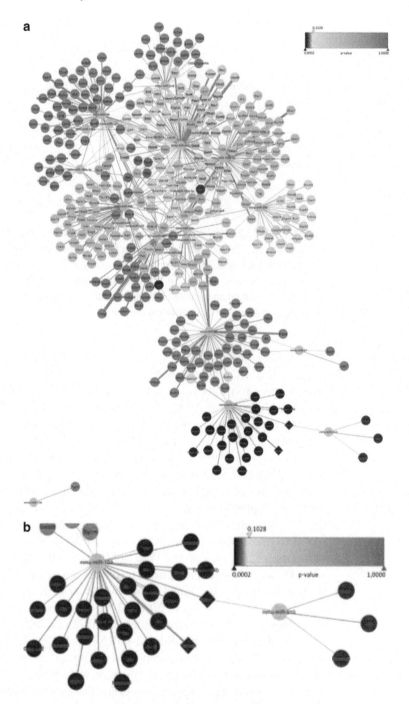

Fig. 3 Pathway dependent visualization of miRNA–mRNA relations with Cytoscape 2.8.3. The enrichments include TargetScan-based miRNA–target predictions as well as mRNA and microRNA expression data. The blue nodes represents microRNAs, whereas the *green*, *yellow* and *red* nodes implying genes. Color intensity (*red*=significant *green*=nonsignificant) indicates the significance of a network area to be overrepresented in a certain pathway or biological context. (**a**) The results for the Neurotrophin signaling pathway are exemplarily depicted. (**b**) Image section of (**a**) The microRNAs 103 and 210 are highly involved in the regulation of the Neurotrophin signaling pathway, corresponding to the overrepresented gene environment

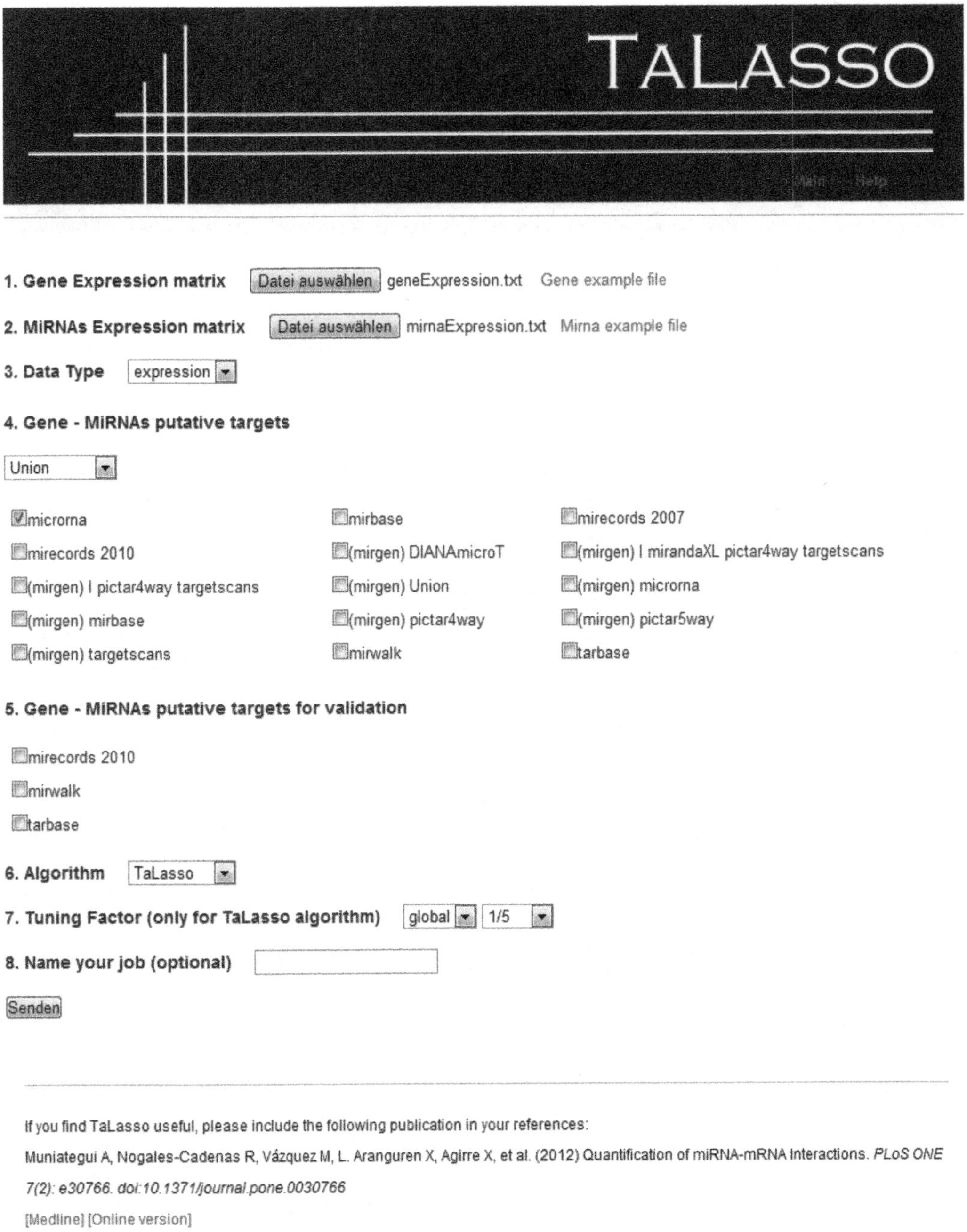

1. Gene Expression matrix   [Datei auswählen] geneExpression.txt   Gene example file

2. MiRNAs Expression matrix   [Datei auswählen] mirnaExpression.txt   Mirna example file

3. Data Type   [expression ▾]

4. Gene - MiRNAs putative targets

[Union ▾]

☑microrna   ☐mirbase   ☐mirecords 2007

☐mirecords 2010   ☐(mirgen) DIANAmicroT   ☐(mirgen) I mirandaXL pictar4way targetscans

☐(mirgen) I pictar4way targetscans ☐(mirgen) Union   ☐(mirgen) microrna

☐(mirgen) mirbase   ☐(mirgen) pictar4way   ☐(mirgen) pictar5way

☐(mirgen) targetscans   ☐mirwalk   ☐tarbase

5. Gene - MiRNAs putative targets for validation

☐mirecords 2010
☐mirwalk
☐tarbase

6. Algorithm   [TaLasso ▾]

7. Tuning Factor (only for TaLasso algorithm)   [global ▾] [1/5 ▾]

8. Name your job (optional)   [   ]

[Senden]

If you find TaLasso useful, please include the following publication in your references:

Muniategui A, Nogales-Cadenas R, Vázquez M, L. Aranguren X, Agirre X, et al. (2012) Quantification of miRNA-mRNA Interactions. *PLoS ONE* 7(2): e30766. doi:10.1371/journal.pone.0030766

[Medline] [Online version]

Fig. 4 Talasso Web interface. Screenshot of the Talasso Web interface, which implements a sequential workflow and different analysis parameters

Therefore, the R environment for statistical computing (http://www.R-project.org/) can be used. This software can be downloaded for free.

Within the R environment, the downloaded assignment matrix ("geneExpression-X_targets.txt") can be imported via $m = as.$ $matrix(read.csv("c:\pathtofile\geneExpression-X\_targets.$

Results for geneExpression-186

Download results files: [Targets] [Genes] [MiRNAs]

Table of results

First   [1]   >>   Last

Gene	miRNA	Score	pValue	Mirecords	Mirwalk	Tarbase
⊞ HECW1 [ENSG00000002746]	⊞ hsa-miR-92	0.10203	0.003072			
⊞ SLC7A2 [ENSG00000003989]	⊞ hsa-miR-15b	0.095237	0.015322			
⊞ EXTL3 [ENSG00000012232]	⊞ hsa-miR-124a	0.065003	4.4876e-05			
⊞ NADK [ENSG00000008130]	⊞ hsa-miR-1	0.037987	0.00103			
⊞ BTBD7 [ENSG00000011114]	⊞ hsa-miR-130a	0.03297	0.099781			
⊞ BTBD7 [ENSG00000011114]	⊞ hsa-miR-30b	0.01116	0.19897			
⊞ CALCR [ENSG00000004948]	⊞ hsa-miR-30b	0.0044773	0.11517			
⊞ PAF1 [ENSG00000006712]	⊞ hsa-miR-17-5p	0.0035542	0.15875			
⊞ LASP1 [ENSG00000002834]	⊞ hsa-miR-1	0.0018174	0.33767	✓	✓	✓
⊞ MAP4K5 [ENSG00000012983]	⊞ hsa-miR-92	0.0016009	0.022455			

Fig. 5 Talasso output. The Talasso result page contains all miRNA–mRNA relationships, which were predicted by the algorithm as well as the score and the *p*-value indicating the significance of the interaction

txt", header = F)). The miRNA identifiers can be imported by the command *colnames = scan ("c:\pathtofile\geneExpression-X_mirna. txt")* and the mRNA identifiers by *rownames = scan ("c:\pathtofile\ geneExpression-X_gene.txt")*. Afterwards, the row and column names of the assignment matrix *m* can be set by *dimnames(m) = lis t(rownames,colnames)*. If you need further help for these commands, just type *? command*, e.g. *? read.csv*. In order to select only those miRNAs and genes in the network that have actual interactions partners, type *m = m[apply(m,1,function(x)sum(x! = 0) > 0),ap ply(m,2,function(x)sum(x! = 0) > 0)]*. The interactions are indicated by the respective *p*-value in the result file which should be logarithmized first. Therefore, type *m[m! = 0] = -log10(m[m! = 0])*. Afterwards, the processed assignment matrix can be exported as xlsx file by *write.xlsx(m,"c:\pathtofile\assignment.xlsx")*. This function requires the package "xlsx" that can be installed by *install. package("xlsx")*. You are now able to import this assignment matrix into graph visualization tools like yEd (http://www.yworks.com). Therefore, open the file by File > Open and choose "assignment matrix". We refer to the yEd user manual for further information on importing and visualizing the network (http://yed.yworks.com/ support/manual/index.html).

3.6 Interpretation of Integrated Expression Data

An important step in the processing of results from integrated miRNA–mRNA analysis is the interpretation of miRNA–target relations. To properly select targets of high significance in the specific biological context and the hypothesis tested for one should

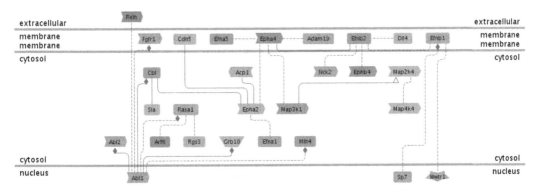

Fig. 6 Enrichment of miRNA targets in signaling pathways based on co-citation. Predicted miRNA targets derived from integrated analysis were analyzed by Genomatix Pathway System algorithm. The results for ephrin are exemplarily depicted. The *red shading* intensity increases with the number of distinct miRNAs, which are inversely correlated to as well as predicted to target the respective gene

acquire quantitative and qualitative parameters on the results from integrated miRNA–mRNA analysis. Parameters include:

1. Number of total targets per miRNA.

2. Number of targeting miRNAs per target.

3. Number of targeted transcription factors per miRNA.

4. Enrichment of targets in canonical pathways, KEGG pathways or pathways based on co-citation (Fig. 6).

5. Enrichment of targets in GO terms, or disease terms etc.

6. Strength of miRNA or mRNA regulation.
   Prioritize and weight parameter based on the biological question you look at and use these information for selection of miRNA–target relations you might want to experimentally validate, e.g., transcription factors which are targeted by even several inversely correlated miRNAs or targets of a specific pathway (*see* **Note 18**).

4   Notes

1. The miScript Reverse Transcription Kit contains 2 buffers. One buffer is the 5× miScript HiSpec Buffer for cDNA preparation. This buffer is used when one intends to do mature miRNA profiling. The other buffer is the 5× miScript HiFlex Buffer for cDNA preparation. This buffer is used when one aims at quantifying mature miRNAs in parallel with precursor miRNAs or mRNAs.

2. A major advantage of the miScript PCR System is the generation of cDNA for both, microRNA and gene expression profiling within the same reaction. On the other hand the miScript PCR

system is very cost intensive. Alternatively, you can quantify the expression of miRNAs and mRNA independently. Latter implies the use of the miScript PCR System for miRNA quantification as described above and the use of gene expression profiling RT kits and SYBR Green kits suited for reverse transcription of mRNAs only.

3. The equipment such as the Affymetrix GeneChip® Fluidics Station 450 and Affymetrix GeneChip® Scanner 3000 7G are cost intensive. At the same time the microarray performance should be standardized according to the MIAME guidelines [40]. Therefore, we recommend to use providers which proceed the arrays (e.g., core facility of the EMBL, Heidelberg).

4. Probes of the GeneChip Gene 1.0 ST Array cap the whole length of the detected genes. This provides a more comprehensive and proper image of gene expression compared to classical 3′ based microarrays. Most 3′-based expression arrays are addicted to transcript's poly-A tails and probes are located to the 3′end of the detected genes. Under certain conditions some genes may not be properly represented on a classical 3′-based expression array such as partially degraded RNA samples, truncated transcripts, or alternative splicing at the 3′end of the gene or polyadenylation sites (www.affymetrix.com).

5. The Gene ST Arrays are the latest of the Affymetrix expression arrays. With the next generation of GeneChip ST Arrays, the GeneChip 2.0 ST Arrays, it is even possible to detect and measure long intergenic RNAs (lincRNA) (www.affymetrix.com).

6. For gene expression analysis profiling, cDNA can be also synthesized by using 1 ng–1 μg. For gene expression analysis of total RNA and 5× First Strand Buffer, 10 mM of each dNTP, 50 μM Hexamere Primer, and 100 U of Moloney Murine Leukemia Virus Reverse Transcriptase; M-MLV RT [H-] (Promega) per reaction. The reaction mix is incubated for 20 min at 21 °C. Then the cDNA synthesis step is performed at 120 min at 48 °C and the reaction is stopped by 2 min incubation at 90 °C. To quantify the gene expression you can use SSo Fast Eva Green Supermix (Bio-Rad Laboratories). The final reaction volume is 10 μl. Use 5 μl SsoFast EvaGreen Supermix, 400 nM forward and 400 nM reverse primers. A total cDNA amount of 50 ng to 50 fg is recommended. The following cycling conditions are verified: After initial activation for 30 s at 98 °C the cycle steps of denaturation for 5 s at 95 °C and annealing/elongation for 20 s at 60 °C are repeated for 39 cycles. Afterwards, a melting curve analysis is performed for each run. The melting curve is generated from 65 to 95 °C with an increment of 5 °C, 5 min.

7. According to the MIQE guidelines [43] normalization of qPCR results is crucial for reducing technical variance. Genes or miRNAs considered as reference should always be evaluated in the specific experimental context prior to applying as normalizers. For evaluation of possible reference genes or microRNAs you can use geNorm [44] or NormFinder [45], both part of the gene expression analysis software suite GenEx (by MultiD Analyses AB).

8. Relative expression changes should be represented as fold-changes (FC) according to the following formula: $FC = 2 - \Delta\Delta Ct$. Including the following calculation steps:

$$\Delta Ct = Ct(\text{target gene}) - Ct(\text{reference gene})$$

$$\Delta\Delta Ct = \Delta Ct(\text{treatment}) - \Delta Ct(\text{control})$$

[46].

9. Hybridization based miRNA microarrays may detect not only mature miRNAs but as well precursor miRNAs. Agilent microRNA microarrays favor the detection of mature miRNAs over precursor forms because of a hairpin-like structure of the probe that sterically reduces the probability of longer oligonucleotide sequences to bind.

10. Agilent's GeneSpring GX software [47] provides statistical tools for visualization and analysis of microarray data such as clustering and principal component analysis, and pathway analysis. For users who are familiar with R language and programming the respective R scripts for statistical analysis and visualization such software is recommended. The R environment is freely available and its use is more flexible in terms of applying different methodologies compared to a given software interface.

11. qPCR assays provide significant advantages over microarrays with respect to sensitivity, dynamic range, and specificity. Probe based qPCR systems such as TaqMan MicroRNA Assays are highly specific and should be favored over DNA intercalating fluorescent dyes. However, probe design is time and cost intensive compared to using universal intercalating dyes.

12. When assay sensitivity is of the utmost importance, or when sample is limiting, a preamplification step using Megaplex™ PreAmp Primers can be added. The preamplification significantly enhances the ability to detect low expressed miRNAs, enabling the generation of a comprehensive expression profile using as little as 1 ng of input total RNA [48]. However, one should keep in mind that preamplification is an additional step which introduces bias.

13. RQ Manager is limited to the simultaneous analysis of 10 cards. This is a disadvantage for larger studies because the analysis has to be partitioned into packages containing no more than 10 cards. Thus, one cannot use for example the automatic threshold function. When having more than 10 cards the automatic adjustment cannot take into account the results of all cards and would calculate different values according to the subset of cards analyzed in parallel. Use fixed threshold option, e.g., threshold at 0.2 instead.

14. Evaluate different normalization methods for your data as inappropriate normalization can strongly affect data quality and the detection of differential expression [43, 44, 49].

15. Consider that in order to calculate reliable relations between miRNA and mRNA expression, an appropriate number of samples in your dataset is necessary. As a rule of thumb, the data should be derived from at least three different conditions or time points with at least three replicates.

16. Application of different methods for the joint analysis of miRNA and mRNA data and comparison of results allow for estimation of the stability of anticorrelated miRNA and mRNA relations over different approaches.

17. Target prediction algorithms do not necessarily contain the latest miRNA annotations according to the latest miRBase version. One needs to assure that miRNAs of interest are in the prediction dataset. Alternatively, one can choose less conservative prediction datasets and use the intersection of results from different prediction algorithms. Results of regression analysis strongly depend on the choice of prediction algorithm and whether one uses the set unity of predictions by different algorithms or the intersection. In general, rather conservative prediction algorithms should be used.

18. Depending on the biological system and hypothesis one wants to test one can implement additional parameters. The priority and weight addressed to each parameter is rather subjective and should be documented carefully to make the individual selection strategy comprehensible.

Acknowledgement

The authors gratefully acknowledge the contribution of Helmut Blum, Stefan Bauersachs, and Stefan Krebs to the conduction of expression profiling. The work was supported by a grant from the German Federal Ministry of Education and Research and the Bavarian State Ministry of Sciences, Research and the Arts.

References

1. Esteller M (2011) Non-coding RNAs in human disease. Nat Rev Genet 12:861–874

2. Fabian MR, Sonenberg N, Filipowicz W (2010) Regulation of mRNA translation and stability by microRNAs. Annu Rev Biochem 79:351–379

3. Krol J, Loedige I, Filipowicz W (2010) The widespread regulation of microRNA biogenesis, function and decay. Nat Rev Genet 11:597–610

4. Westholm JO, Lai EC (2011) Mirtrons: microRNA biogenesis via splicing. Biochimie 93:1897–1904

5. Kim VN, Han J, Siomi MC (2009) Biogenesis of small RNAs in animals. Nat Rev Mol Cell Biol 10:126–139

6. Curtis HJ, Sibley CR, Wood MJA (2012) Mirtrons, an emerging class of atypical miRNA. Wiley Interdiscip Rev RNA 3:617–632

7. Djuranovic S, Nahvi A, Green R (2012) miRNA-mediated gene silencing by translational repression followed by mRNA deadenylation and decay. Science 336:237–240

8. Bartel DP (2009) MicroRNAs: target recognition and regulatory functions. Cell 136:215–233

9. Chi SW, Hannon GJ, Darnell RB (2012) An alternative mode of microRNA target recognition. Nat Struct Mol Biol 19:321–327

10. Kumar A, Wong AK, Tizard ML et al (2012) miRNA_Targets: a database for miRNA target predictions in coding and non-coding regions of mRNAs. Genomics 100:352–356

11. Forman JJ, Legesse-Miller A, Coller HA (2008) A search for conserved sequences in coding regions reveals that the let-7 microRNA targets Dicer within its coding sequence. Proc Natl Acad Sci U S A 105:14879–14884

12. Guo H, Ingolia NT, Weissman JS et al (2010) Mammalian microRNAs predominantly act to decrease target mRNA levels. Nature 466:835–840

13. Lee S, Vasudevan S (2013) Post-transcriptional stimulation of gene expression by microRNAs. Adv Exp Med Biol 768:97–126

14. Vasudevan S, Tong Y, Steitz JA (2007) Switching from repression to activation: microRNAs can up-regulate translation. Science 318:1931–1934

15. Enerly E, Steinfeld I, Kleivi K et al (2011) miRNA–mRNA integrated analysis reveals roles for miRNAs in primary breast tumors. PLoS One 6:e16915

16. Chi SW, Zang JB, Mele A et al (2009) Argonaute HITS-CLIP decodes microRNA–mRNA interaction maps. Nature 460:479–486

17. Muniategui A, Pey J, Planes FJ et al (2013) Joint analysis of miRNA and mRNA expression data. Brief Bioinform 14:263–278

18. Affymetrix I GeneChip® Gene 1.0 ST Array System for Human, Mouse and Rat. A simple and affordable solution for advanced gene-level expression profiling. Data sheet. http://affy.arabidopsis.info/documents/gene_1_0_st_datasheet.pdf

19. Saeed AI, Sharov V, White J et al (2003) TM4: a free, open-source system for microarray data management and analysis. BioTechniques 34:374–378

20. Howe E, Holton K, Nair S et al (2010) MeV: MultiExperiment viewer. In: Ochs MF, Casagrande JT, Davuluri RV (eds) Biomedical informatics for cancer research. Springer, New York, USA, pp 267–277

21. Lewis BP, Burge CB, Bartel DP (2005) Conserved seed pairing, often flanked by adenosines, indicates that thousands of human genes are microRNA targets. Cell 120:15–20

22. Friedman RC, Farh KK, Burge CB et al (2009) Most mammalian mRNAs are conserved targets of microRNAs. Genome Res 19:92–105

23. Betel D, Wilson M, Gabow A et al (2008) The microRNA.org resource: targets and expression. Nucleic Acids Res 36(Database issue):D149–D153

24. Enright AJ, John B, Gaul U et al (2003) MicroRNA targets in Drosophila. Genome Biol 5:R1

25. Lu Y, Zhou Y, Qu W et al (2011) A Lasso regression model for the construction of microRNA-target regulatory networks. Bioinformatics 27:2406–2413

26. Ashburner M, Ball CA, Blake JA et al (2000) Gene ontology: tool for the unification of biology. The Gene Ontology Consortium. Nat Genet 25:25–29

27. Schaefer CF, Anthony K, Krupa S et al (2009) PID: the pathway interaction database. Nucleic Acids Res 37(Database issue):D674–D679

28. Qiagen (2012) miRNeasy Mini Handbook – (EN). For purification of total RNA, including miRNA, from animal and human cells and tissues. http://www.qiagen.com/Knowledge-and-Support/Resource-Center/Resource-Search/

29. Hartung T, Balls M, Bardouille C et al (2002) Good cell culture practice. ECVAM good cell culture practice task force report 1. Altern Lab Anim 30:407–414

30. Wegener J, Keese CR, Giaever I (2000) Electric cell-substrate impedance sensing (ECIS) as a

noninvasive means to monitor the kinetics of cell spreading to artificial surfaces. Exp Cell Res 259:158–166

31. Agilent Technologies DNA, RNA, protein and cell analysis. Agilent 2100 bioanalyzer. Lab-on-a-chip technology. http://www.icmb.utexas.edu/core/DNA/Information_Sheets/Bioanalyzer/bioanalyzer.pdf

32. Qiagen (2011) miScript PCR System Handbook. For real-time PCR analysis of microRNA using SYBR Green detection. http://www.qiagen.com/Knowledge-and-Support/Resource-Center/Resource-Search/?q=miScript+PCR+System+Handbook%3b&l=en%3b

33. Dufva M (2009) Introduction to microarray technology. Methods Mol Biol 529:1–22

34. Affymetrix I GeneChip® Whole Transcript (WT) Sense Target Labeling Assay. User Manual. http://www.affymetrix.com/estore/browse/products.jsp?categoryIdClicked=&productId=131467#1_3

35. Affymetrix I (2008) GeneChip® Expression Wash, Stain and Scan User Manual. For Cartridge Arrays. http://www.affymetrix.com/estore/browse/products.jsp?categoryIdClicked=&productId=131467#1_3

36. Risso D, Massa MS, Chiogna M et al (2009) A modified LOESS normalization applied to microRNA arrays: a comparative evaluation. Bioinformatics 25:2685–2691

37. Meyer SU, Kaiser S, Wagner C et al (2012) Profound effect of profiling platform and normalization strategy on detection of differentially expressed microRNAs—a comparative study. PLoS One 7:e38946

38. Brazma A, Parkinson H, Sarkans U et al (2003) ArrayExpress—a public repository for microarray gene expression data at the EBI. Nucleic Acids Res 31:68–71

39. Edgar R, Domrachev M, Lash AE (2002) Gene expression omnibus: NCBI gene expression and hybridization array data repository. Nucleic Acids Res 30:207–210

40. Brazma A, Hingamp P, Quackenbush J et al (2001) Minimum information about a microarray experiment (MIAME)-toward standards for microarray data. Nat Genet 29:365–371

41. Tusher VG, Tibshirani R, Chu G (2001) Significance analysis of microarrays applied to the ionizing radiation response. Proc Natl Acad Sci U S A 98:5116–5121

42. van der Auwera I, Limame R, van Dam P et al (2010) Integrated miRNA and mRNA expression profiling of the inflammatory breast cancer subtype. Br J Cancer 103:532–541

43. Bustin SA, Benes V, Garson JA et al (2009) The MIQE guidelines: minimum information for publication of quantitative real-time PCR experiments. Clin Chem 55:611–622

44. Vandesompele J, Preter K, de Pattyn F et al (2002) Accurate normalization of real-time quantitative RT-PCR data by geometric averaging of multiple internal control genes. Genome Biol 3:RESEARCH0034

45. Andersen CL, Jensen JL, Ørntoft TF (2004) Normalization of real-time quantitative reverse transcription-PCR data: a model-based variance estimation approach to identify genes suited for normalization, applied to bladder and colon cancer data sets. Cancer Res 64:5245–5250

46. Livak KJ, Schmittgen TD (2001) Analysis of relative gene expression data using real-time quantitative PCR and the 2(-Delta Delta C(T)) Method. Methods 25:402–408

47. Agilent Technologies GeneSpring GX 9.0 QuickStartGuide. http://www.chem.agilent.com/library/usermanuals/Public/GeneSpringGX9_QuickStartGuide.pdf

48. Applied Biosystems Megaplex™ Pools For microRNA Expression Analysis. Protocol. http://www3.appliedbiosystems.com/cms/groups/mcb_support/documents/general-documents/cms_053965.pdf

49. Meyer SU, Pfaffl MW, Ulbrich SE (2010) Normalization strategies for microRNA profiling experiments: a "normal" way to a hidden layer of complexity? Biotechnol Lett 32:1777–1788

Chapter 16

Clinical Applications Using Digital PCR

Francisco Bizouarn

Abstract

Molecular diagnostics and disease-specific tailored treatments are now being introduced to patients at many hospitals and clinics throughout the world (Strain and Richman, Curr Opin HIV AIDS 8:106–110, 2013) and becoming prevalent in the nonscientific literature. Instead of generically using a "one treatment fits all" approach that may have varying levels of effectiveness to different patients, patient-specific molecular profiling based on the genetic makeup of the disease and/or a more accurate pathogen titer could provide more effective treatments with fewer unwanted side effects.

One commonly known example of this scenario is epidermal growth factor receptor (EGFR). EGFR is upregulated in many cancers, including many lung and colorectal cancers. Commonly used treatments for these include the receptor blockers cetuximab or panitumumab and tyrosine kinase inhibitors erlotinib or gefitinib. These agents are effective at reducing out-of-control cell cycling and tumor proliferation, but only if downstream signaling kinases and phosphatases are not mutated. Known oncogenes such as BRAF V600E and KRAS G12/13 that are constitutively activated render these treatments ineffective. The use of known ineffective drugs and treatments can thus be avoided reducing time to more effective treatments, reducing cost, and increasing patient well-being.

Although digital PCR is for all practical purposes a "new" technology, there is already tremendous interest in its potential for the clinical diagnostics arena. Specificity of the information acquired, accuracy of results, time to results, and cost per sample analyzed are making dPCR an attractive tool for this field. Three areas where dPCR will have a noticeable impact are pathogen/viral detection and quantitation, copy number variations, and rare mutation detection and abundance, but it will inevitably expand from these as the technology becomes more and more prevalent.

This chapter discusses digital PCR assay optimization and validation, pathogen/viral detection and quantitation, copy number variation, and rare mutation abundance assays. The sample methods described below utilize the QX100/QX200 methodologies, but with the exception of reaction sub-partitioning (dependent on the instrumentation used) most other parameters remain the same.

Key words Digital PCR assay optimization and validation, Copy number variations, Mutation abundance, Quantitation, Viral load, Pathogen detection

1 Introduction

As with any assay, proper assay validation and sometimes optimization when using digital PCR are essential. Although most validated qPCR assays work right from the start, even when combined in duplex, it is always prudent to do due diligence and confirm that

Roberto Biassoni and Alessandro Raso (eds.), *Quantitative Real-Time PCR: Methods and Protocols*, Methods in Molecular Biology, vol. 1160, DOI 10.1007/978-1-4939-0733-5_16, © Springer Science+Business Media New York 2014

the assay works as intended prior to initiating any serious work. It is highly recommended to follow the "The Digital MIQE Guidelines" manuscript [1].

Three steps need to be fulfilled in the dPCR assay optimization and validation:

Thermal optimization

Validation of accuracy

Validation of specificity

Here we present typical protocols where digital PCR produces unrivaled data compared to other methodologies.

1.1 Pathogen/Viral Detection and Quantitation

The proper detection and quantization of pathogen and viral particles can be quite challenging with clinical samples, as they are obtained from many parts of the body, in forms ranging from fluid to solid, concentrated to very dilute, and from organisms that generally tend to have high random mutation rates. PCR and qPCR, with their high potential to detect and amplify extremely low levels of target molecules, are obvious solutions of choice for clinical sample analysis, but their accuracy can sometimes be compromised. Sample extraction and removal of inhibitory molecules can be difficult and can dramatically affect PCR and qPCR results, usually resulting in an under-representation of the actual amount of target molecule present. In certain cases, this inhibition can lead to a weak amplification of qPCR profiles leading to false-negative calls for positive samples. Random point mutations in the primer annealing sites can significantly disrupt initial amplification of low-level target molecules delaying the appearance of the amplification profiles and subsequent Cq's. A single base mismatch in the 3' end of the primer can easily delay the amplification by five cycles. A mismatch at the penultimate base can cause a delay of three cycles and a delay of one to two cycles (these values vary as a function of reaction supermix used and primer concentration and base composition). Due to the exponential facet of qPCR, these delays can cause large errors in quantitation and potentially detection.

Digital PCR can help solve these issues, as it is a digital analysis of thousands of individual sub-partitioned PCR assays. Quantitative values generated are absolute and independent of other samples that may or may not be accurate. Not depending on an external standard curve that may or may not represent adequately the sample being analyzed makes dPCR very attractive [2–5]. Assays can be run in singleplex for direct detection and quantitation or can be run in duplex for a second target of interest or for the monitoring of an endogenous or an exogenous reference (Fig. 1).

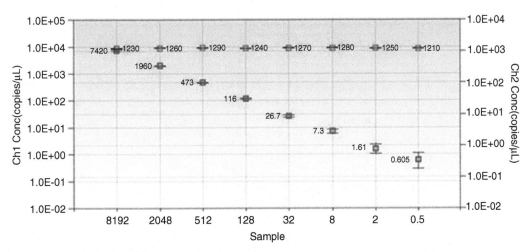

Fig. 1 Example of a duplex titration assay. *S. aureus* was titered down in a fourfold dilution series from 8,192 copies per ml to approximately 0.5 copies per μl (164,000 to 10 copies per 20 μl reaction) in the presence of 1,250 copies per ml of human RPP30 (25,000 copies per 20 μl reaction). Each sample represents a single reaction well

1.2 Copy Number Variation Analysis

It encompasses disorders that vary from simple gene deletions and trisomies to inversions and translocations. The effects associated with these decrease or increase in the number of copies if a specific gene or sets of genes (up to chromosomes) vary from no apparent affect to enhancements (as in tolerance to pesticides, herbicides, parasites, and drought in plants), cancers (MYC, AKT2, EGFR, etc.), and premature deaths (as with many trisomies). Current methods of quantifying copy number variations (CNVs) are FISH, next-generation sequencing, qPCR micro arrays, and Southern blotting. Most of these are laborious and expensive and generate results of varying accuracy.

CNV analysis requires two measurement steps. The first is to accurately determine the amount of starting material to be used in the quantitation of the targets of interest. This is typically done using one or more reference or control genes, of known copy number, that are unaffected by the disease, distal from the replicated site, or on a different chromosome (EIF2C1, AP3B1, RPP30, etc.). From this information one can determine the initial number of genomes being analyzed in the assay and use it as a normalization factor when quantitating the target genes of interest.

The second step is to accurately quantitate the target genes of interest.

Digital PCR provides a fast, cost-effective, and high-resolution tool for CNV analysis [6–9] (Fig. 2).

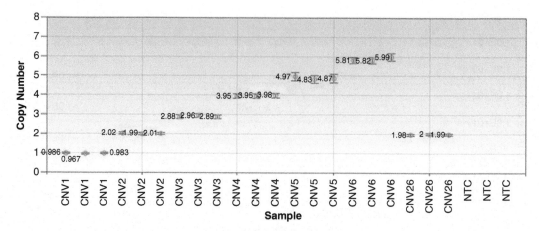

Fig. 2 MGPRX1 copy number analysis of seven different Coriell cell lines with copy number variations from 1 to 6 and an NTC, each run in triplicate. Each result represents a single reaction well with a duplex PCR detecting the presence of MGPRX1 using a FAM-labeled hydrolysis probe and RPP30 using a HEX-labeled hydrolysis probe (used as the reference). Data generated using QX100 Droplet Digital PCR system

1.3 Rare Mutation Detection

Many diseases carry in their genetic profile point or small-mutated sequences that confer attributes to the cell that may or may not be desirable. Detecting a mutation on one allele of a homogeneous sample, where the distribution is 100 % WT, 100 % Mut, or 50 % WT 50 % Mut, is relatively straightforward. Of greater interest is the ability to detect these mutations in a very large background of normal cells or tissues. Instead of monitoring the diseased tissue directly, less invasive procedures and more readily accessible sources (plasma, serum, urine, etc.) may potentially be used for testing and monitoring.

The technical challenge, when using amplification-based methods, lies in the fact that when amplifying a target in these samples, both wild-type and mutant variants will amplify at approximately the same rate making the mutant variant difficult to detect. This is analogous to the proverbial needle in a haystack description. If one uses primer-specific amplification strategies (primer anchoring, nested assays, modified nucleotides) detection capability can improve slightly, but these assays are very temperamental to experimental condition variations such as protocols and the presence of inhibitors. Historically, rare mutation detection (RMD) assays were able to detect levels of 5–10 % mutations with few exceptional cases at levels near or better than 1 %.

In digital PCR the massive sub-partitioning of a PCR assay creates a synthetic enrichment effect that can be utilized to dramatically enhance the detection capability of these rare mutations at very low levels [8, 10].

If a bulk 20 µl PCR reaction contains 40 mutant target molecules in a background of 40,000 wild-type molecules, the ratio of Mut to WT is 1 in 1,000. This will be difficult to detect even after

Bulk Sample - 20 µL

Partitioned Sample – 20,000 × 1nL

40,000 target molecule A
40 target molecule B

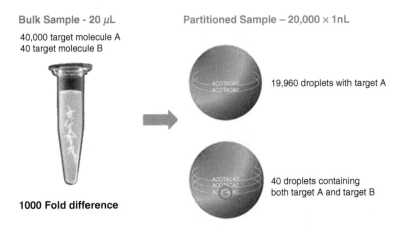

19,960 droplets with target A

40 droplets containing
both target A and target B

1000 Fold difference

Fig. 3 Sub-partitioning of a PCR reaction and subsequent analysis of the partitions cause a synthetic enrichment effect that permits detection of rare or low-abundance mutations from within a large population of normal sample. Instead of looking for something very diluted in a large volume, we look at many small volumes where proportionally the mutation of interest is at a higher abundance

amplification, as this 1,000-fold difference will be maintained during the amplification process. If we partition the 20 µl into 20,000 sub-partitions, the average partition will contain two WT molecules and 40 of these partitions will also contain one mutant with a ratio of 1 in 3. When amplified this ratio is maintained. Each sub-partition is interrogated individually, and those with a mutant molecule become easily recognizable. In real-life partitioning experiments, the distribution of wild-type molecules within the partition population will follow a Poisson distribution, with each having between 0 and possibly 7 or 8 molecules within. Even at these higher levels, a mutant target molecule will be easily detectable as it will at worse be at a 1-to-8 or 1-to-9 ratios (Fig. 3).

In classical qPCR random mutation abundance genotyping hydrolysis assays, two probes are designed to land on the designated target gene area with as little difference between them as a single base detecting either WT or Mut alleles. The probe targeting the wild-type sequence is typically labeled with a HEX fluorophore, and the mutant-hybridizing probe is typically labeled with a FAM fluorophore. In digital PCR technique, both assays can be run simultaneously in the same tube using qPCR-based probe techniques. These assays are easy to set up, similar to qPCR assays, require little hands on time, and give astonishing results (Fig. 4).

2   Materials

1. Restriction enzymes and associated buffers (when needed).

2. Purified sample: Sample to be tested should be as clean and inhibitor free as possible.

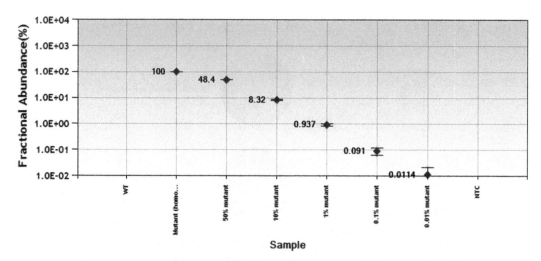

Fig. 4 Fractional abundance of BRAF V600E mutant. Merged data of three wells representing wild type, 100, 50, 10, 1, 0.1, and 0.01 % mutant, as well as corresponding NTC

3. Primer- and FAM-labeled probe mix. Primer concentration 18 μM and probe concentration 5 μM, both at 20× (900 nM primer, 250 nM probe at 1× final) for the target of interest assay.

4. Primer- and HEX-labeled probe mix: Primer concentration 18 μM, probe concentration 5 μM, both at 20× (900 nM primer, 250 nM probe at 1× final) for reference assay.

5. Digital PCR instrumentation (QX100 or QX200 from Bio-Rad Laboratories for this example) and associated instrument operation and data analysis software.

6. Digital PCR reagents (QX100 and QX200):

   (a) Digital PCR Supermix if using DNA (2× mix).

   (b) One-Step RT-ddPCR Kit for probes if using RNA (2× mix).

   (c) Droplet generator oil.

   (d) Droplet reader oil.

   (e) Droplet generator cartridge and gaskets.

7. 96-well plate.

8. Heat seal foil.

9. Heat sealer.

10. Single- and multichannel (P8×20 μl, P8×50 μl, P8×200 μl) micropipettes.

11. Pipet tips with aerosol barriers.

12. 1.5 ml tubes.

3 Methods

3.1 Thermal
Optimization

1. In a 1.5 ml tube prepare sufficient reaction mix for nine gradient PCR preparations. Eight will be used for the gradient run. One is extra (*see* **Note 1**).

2. 90 µl of 2× digital PCR supermix.

3. 9 µl 20× primer and probe mix of first (target) assay (*see* **Note 2**).

4. 9 µl 20× primer and probe mix of second (reference) assay (*see* **Note 2**).

5. Xµl of sample; amount to be used should correspond to between 2,000 and 50,000 copies per reaction (10,000–20,000 is ideal) (*see* **Note 3**).

6. Υµl of water required for total reaction volume of 180 µl.

7. Thoroughly mix the reaction (*see* **Note 4**).

8. PCR reaction sub-partitioning (on QX100 and QX200) (*see* **Note 5**).

9. Insert a new ddPCR cartridge into cartridge holder (*see* **Note 6**).

10. Transfer 20 µl of gradient PCR reaction mix into the eight sample wells of the droplet generator cartridge (Fig. 5) (*see* **Note 7**).

11. Deposit 70 µl of droplet generator oil into the eight oil wells of the droplet generator cartridge (Fig. 5).

12. Place new rubber gasket onto cartridge holder (Fig. 6).

13. Place the holder in droplet generator, and close lid (*see* **Note 8**).

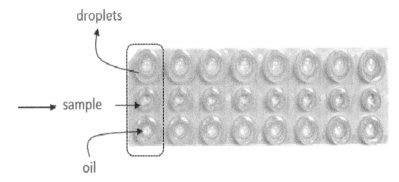

Fig. 5 Disposable droplet-generating cartridge used on the QX100/QX200 Droplet Digital PCR system. Cartridge has independent eight sets of microfluidic channels that permit the simultaneous partitioning of eight different samples. Typically 20 µl of reaction mix is deposited into the center wells, and 70 µl of droplet-generating oil is deposited into the oil wells. Post-processing, 40 µl of a droplet/oil mixture is recovered from the droplet chamber

Fig. 6 The droplet generator cartridge covered with gasket inserted into the droplet generator. The processing of eight samples takes about 2 min. Post-processing, the droplets are transferred to a 96-well plate

14. When the lid opens, sub-partitioned droplets are ready. Gently transfer droplets into a 96-well plate using a P8×50 μl multichannel micropipette (*see* **Note 9**).

15. Discard used cartridge and gasket.

16. When finished loading the plate, seal using foil film.

3.2 **Amplification**   Load the sealed 96-well reaction plate into a gradient capable thermocycler noting plate orientation.

1. Program protocol as follows: 95 °C for 10 min followed by 45 cycles of 30 s at 95 °C and 1 min at annealing temp, where higher end of gradient is 10° above design temp and lower end of gradient is 5° below.

2. When cycling is complete, transfer plate to the droplet reader. The droplet reader collects the droplets from each well sequentially and aligns them sequentially on a flow of reader oil. The proplets are interrogated for the presence or the absence of fluorescence of FAM and HEX fluorescence (Fig. 7). These results are presented in the results and data analysis section on the operating software.

3. ddPCR gradient data analysis: Set data analysis to view eight gradient reactions together (Fig. 8) (*see* **Note 10**).

4. Select a temperature that provides good separation between positive and negative droplets (or partitions) (Fig. 8). Conditions for G06 are probably ideal, but F06 and H06 are also good (*see* **Notes 10** and **11**).

5. As long as there is a clear separation between the two clusters, for a similar assay, quantitative results will vary little as a function of different temperatures (Fig. 9).

6. If running a duplex assay, select reaction conditions that are optimal for both assays; conditions used in D01, E01, and F01 are optimal (Fig. 10) (*see* **Notes 12** and **13**).

7. As with singleplex reactions, as long as the two groups of droplets (positive and negative) are clearly differentiated, results will remain reasonably similar (Fig. 11).

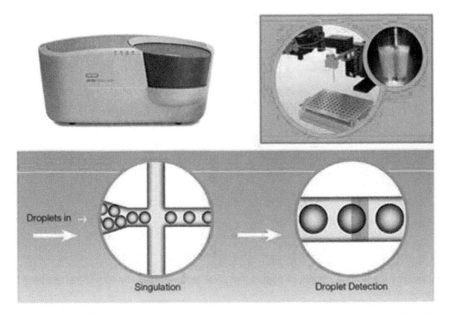

Fig. 7 QX100/QX200 Droplet Reader. Once cycled, the droplets are collected from a well of the 96-well plate, aligned in single file suspended in fluidic reader oil, and individually interrogated in the FAM and HEX fluorescent ranges. Collected data is forwarded to a computer for further analysis

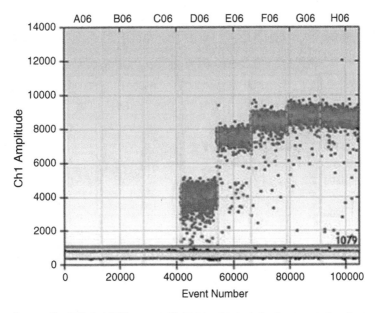

Fig. 8 Example of a gradient digital PCR assay. Eight identical replicates were simultaneously run under slightly different annealing conditions (55–65 °C gradient) to determine optimal annealing temperature for the specific assay

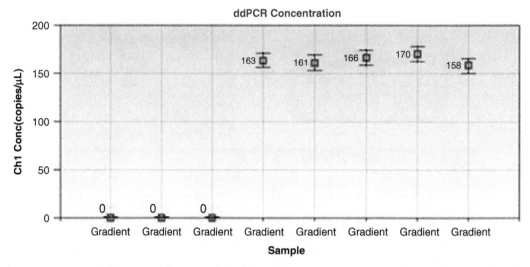

Fig. 9 Numerical concentration results in copies per µl of input reaction mix from the gradient experiment presented in Fig. 7

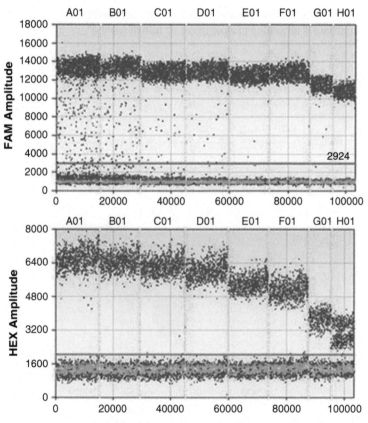

Fig. 10 Duplex thermal gradient (annealing/extension step) from 55 to 65 °C across wells A1 through H1. Human GAPDH was amplified using a FAM-labeled assay, and *S. aureus* was amplified using a HEX-labeled assay. Optimal assay conditions are wells that show good separation of positive and negative clusters (C01, D01, E01, and F01) and little rain (partitions where nonspecific products were formed A01 and B01). Each sample represents a single well

3.3 Accuracy and Specificity Validation

Validation requires a series of tests to confirm that it meets set expectations. At a minimum, the following should be verified:

1. Verify specificity for the molecular target of interest preferably using sequencing.

2. Verify precision by running multiple replicates of samples spanning the dynamic range of interest.

3. Verify sensitivity by carefully serially diluting template (with a true negative control) to levels where it can no longer be differentiated from true-negative controls (see **Note 14**).

4. Establish the limit of detection using multiple true-negative controls.

5. Verify the effect potential inhibitors present in the sample by titering up its presence in a control positive reaction.

3.4 DNA Extraction for Subsequent Validation Analysis

1. Pipet and discard the bottom oil phase of the well.

2. Add 10 µl of TE.

3. In a fume hood, add 40 µl of chloroform.

4. Pipet up and down five times at a setting of 25 µl volume.

5. Pipet out the entire volume into 1.5 ml tube, and cap the tube.

6. Vortex at maximum speed for 1 min.

7. Centrifuge at $15,500 \times g$ for 10 min.

8. Remove the upper phase carefully by pipetting, avoiding the chloroform phase, and transfer to another 1.5 ml tube.

9. Dispose of the chloroform phase appropriately.

10. Sample can be further prepped for sequencing using spin micro columns.

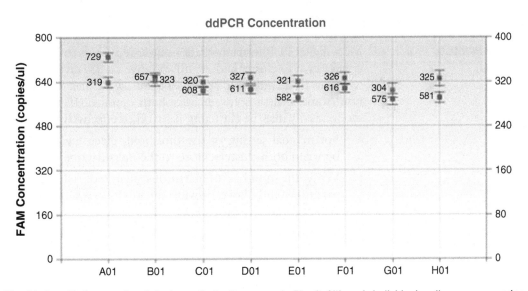

Fig. 11 Quantitative results of duplex optimization assay in Fig. 9. Although individual wells were run under different conditions, results are very similar

3.5 Pathogen/Viral Detection and Quantitation

The previous points of setup present all the steps needed to optimize ddPCR experiments and have to be followed before attempting to analyze any samples.

1. PCR reaction preparation: If the sample to be used is very concentrated, consider digesting it with a restriction enzyme prior to preparing the PCR reaction (*see* **Note 15**). DNA concentrations above 60 ng per 20 μl PCR reaction should be digested (*see* **Note 16**). Once digested, concentrations of 1 μg per 20 μl PCR reaction (and sometimes more) can be used.

2. If using a low-salt buffer for the digest, the sample will most likely need not be prepped (column or precipitated) prior to use in the PCR reaction (*see* **Note 17**).

3. In a 1.5 ml tube or in a well of a 96-well plate mix reaction components: 10 μl of 2× digital PCR supermix, 1 μl 20× primer and probe mix of first (target) assay, 1 μl 20× primer and probe mix of second target (reference) assay, Xμl of sample, and Yμl of water required for total reaction volume of 20 μl. Thoroughly mix the reaction. This can be done by pipet-mixing or using a vortex (*see* **Note 4**).

3.6 PCR Reaction Sub-partitioning (on QX100 and QX200)

Follow all the steps from **steps 9** to **16** in Subheading 3.1 (*see* **Notes 5–9**).

3.7 Amplification

1. Load the sealed 96-well reaction plate into a thermocycler.

2. Program protocol as follows: 95 °C for 10 min followed by 45 cycles of 30 s at 95 °C and 1 min at the selected annealing temperature (*see* **step 4**, Subheading 3.2).

3. When cycling is complete, transfer plate to the droplet reader (*see* **step 2**, Subheading 3.2) (Fig. 7).

3.8 ddPCR Data Analysis

1. Most digital PCR instrumentations include analysis software that either can auto-cluster and identify positive vs. negative events or will allow for a "threshold" to be set manually for their differentiation (some analysis software offer both options). If using auto modes, verify that the clustering has worked effectively.

   • For manual setting of the threshold, place the threshold between both clusters, close to the negative one (Fig. 12).

   • Verify the number of partitions analyzed. 10,000 sub-partitions and above provide quantitative results that are very accurate (*see* **Note 18**).

   • Software will subsequently calculate concentrations in the form of copies per microliter of PCR reaction. If a 20 μl reaction was used, multiply the value(s) by 20 to get the total number for a complete reaction (Fig. 13) (*see* **Notes 19** and **20**).

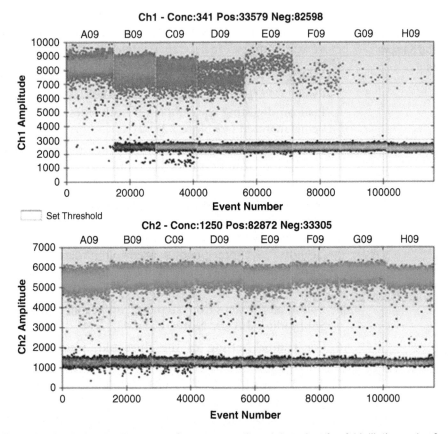

Fig. 12 Example of a duplex titration assay. *S. aureus* was titered down in a fourfold dilution series from 8,192 copies per μl to approximately 0.5 copies per μl (164,000 to 10 copies per 20 μl reaction) in the presence of 1,250 copies per μl of human RPP30 (25,000 copies per 20 μl reaction). Each sample represents a single reaction well

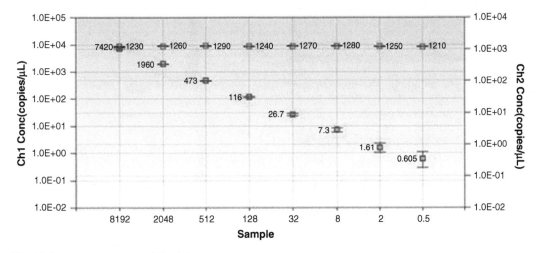

Fig. 13 Quantitative results of Fig. 15

3.9 Copy Number Variation Experimental Planning and Sample Preparation

The steps from Subheading 3.1 to **step 10** of Subheading 3.4 are needed to optimize ddPCR experiments and have to be followed before attempting to analyze any samples.

1. Select a reference target that is stable and away from the suspected area of repeated targets of interest (*see* **Note 21**).

2. Plan on digesting the sample on both sides of the amplicons being analyzed (*see* **Note 22**).

3. Samples may be analyzed in both digested and undigested stares for proximity studies (*see* **Note 23**).

4. Digest purified DNA according to the enzyme manufacturer's recommendations. If sample is digested in a low-salt buffer, cleanup may not be necessary and the digested sample can be used directly in the dPCR reaction. If using a high-salt buffer, cleanup DNA (precipitation, prep column, etc.) or use a small amount of material if possible (1 or 2 µl of digestion product in 20 µl reaction).

5. Follow all the steps from **step 3** of Subheading 3.5 to **step 1** of Subheading 3.8 (*see* **Notes 1** and **4**).

6. For manual setting of the threshold, place the threshold between both clusters, close to the negative one (Fig. 14).

7. The 2D view should show nice orthogonal distribution of the sub-assays into four clusters; if not, reaction conditions need to be tweaked (annealing temp, primer concentration, etc.) (Fig. 15).

8. System software will subsequently compare the levels of both targets (reference and interest) and call out a copy number per genome (*see* **Notes 24–26**) (Fig. 16).

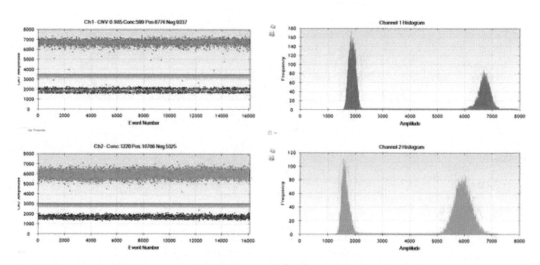

Fig. 14 Copy number variation assay (single well). If manual threshold setting is required, set them manually slightly above negative sub-partitions (*red lines*)

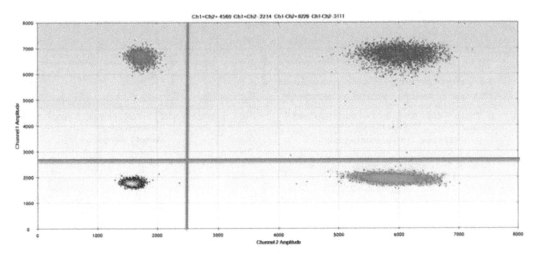

Fig. 15 2D plot of copy number variation assay (single well). If manual threshold setting is required, set cross-hairs close to negative sub-partitions (*red lines*)

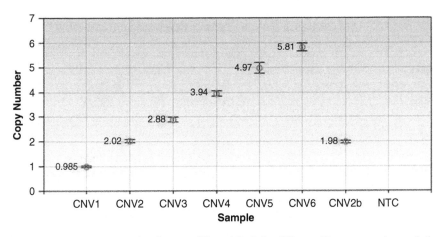

Fig. 16 MGPRX1 copy number analysis of seven different Coriell cell lines with copy number variations from 1 to 6 and an NTC. Each result represents a single reaction well with a duplex PCR detecting the presence of MGPRX1 using a FAM-labeled hydrolysis probe and RPP30 using a HEX-labeled hydrolysis probe (used as the reference)

3.10 Rare Mutation Detection

The steps from Subheading 3.1 to **step 10** of Subheading 3.4 are needed to optimize ddPCR experiments and have to be followed before attempting to analyze any samples.

1. Follow all the steps from **step 3** of Subheading 3.5 to **step 3** of Subheading 3.7 (*see* **Notes 4, 27–29**).

2. Mutation abundance data often looks different from quantitative or CNV results. There are often what appear to be multiple baselines of negative subreactions in the temporal (event) plots (Fig. 17). Temporal or event plot should not be used for analysis! Analysis must be performed using the 2D

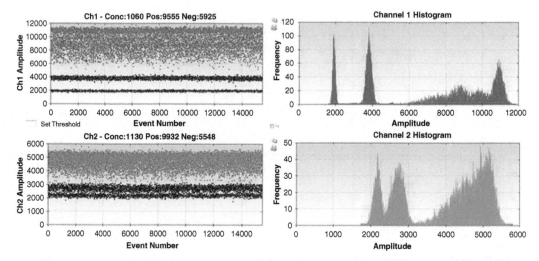

Fig. 17 Temporal (event) plot of a rare mutation abundance assay. This view should not be used for analysis as slight cross-reactivity of WT and Mut probes can make calling events difficult

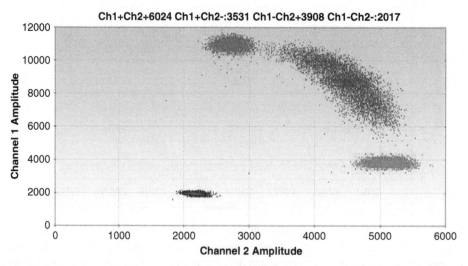

Fig. 18 2D plot of a rare mutation abundance assay. Preferred method for partition analysis. *Grey dots* are negative partitions for both WT and Mut, *blue* are positive for Mut only, *green* are positive for WT only, and *red* are positive for both WT and Mut genes

plots (Fig. 18). Although software algorithms should be capable of clustering the partitioned groups, depending on the level of cross-reactivity of the probes, it may be necessary to set them manually (*see* **Note 30**).

3. Using a lasso tool, loop the clusters individually. Use the mutant and WT control samples for positional reference. The four clusters are composed of three typical clusters and a fourth "fanned-out" cluster (Fig. 19). The bottom left cluster are the empty sub-partitions, the upper left (in blue) are FAM positive,

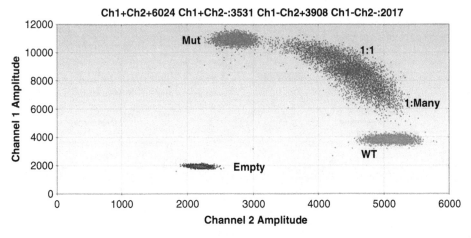

Fig. 19 Fanning effect of partitions that contain both WT and Mut molecules. As the abundance of WT molecules increases vs. the rare mutant molecule the partition's fluorescent profile migrates down towards the WT cluster

the lower right are the HEX positive, and the fanned ones are the double positive (both FAM and HEX) (Fig. 19).

4. As both WT and mutant targets are quantified, a fractional abundance of mutant molecules (within the total) can be determined for each sample (Figs. 20 and 21).

4   Notes

1. Normally a sample would be analyzed in a single well, but for a gradient run eight wells are typically used for each condition used, as many thermocyclers have blocks with gradients that span eight temperatures across eight wells. An extra reaction's worth of mix is prepared for the gradient test as often; when preparing these mixes, one runs out of mix at the last well one intends to pipet (the famous fudge factor).

2. Typical primer and probe concentrations used are 900 nM for primers and 250 nM for probes. As with PCR and qPCR these concentrations can be modified and reduced. Proper validation at the new concentrations is required.

3. The amount of sample used for the gradient reaction can be estimated using optical density. The goal is to have an amount of starting material where the target of interest will fall within the dynamic range of the instrument (1–100,000 copies with 10,000–20,000 optimal).

4. The reaction mix should be as uniform as possible such that the only variable between wells is the temperature. Proper mixing is critical in digital PCR, as once the main mix is partitioned, there will be no mixing with reagent components

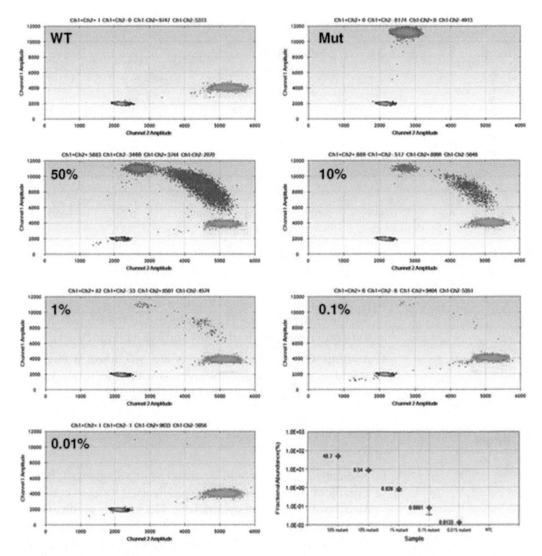

Fig. 20 Titration of mutated BRAF V600E vs. normal BRAF at 50, 10, 1, 0.1, and 0.01 % and NTCs in duplicate. *Bottom right panel* is fractional abundance chart

outside that partition. If one side (or top or bottom) of the tube has more concentrated reaction components and the tube is not mixed, there will be improper reagent conditions in many of the wells. This will cause the assay to fail and also nullify Poisson distribution-expected patterns.

5. If using an instrument other than the QX100 and QX200, follow the manufacturer's recommended protocol for partitioning.

6. Regardless of the instrument manufacturer partitioning system used, note that consumables cannot be reused. Microfluidic channels and microwells cannot be effectively cleaned, and

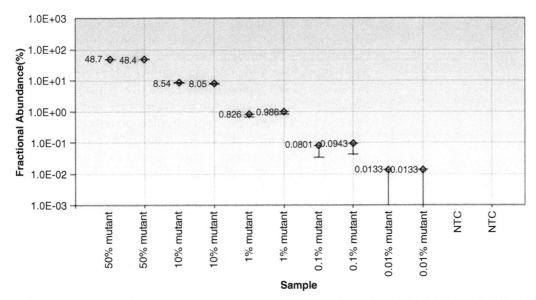

Fig. 21 Fractional abundance of mutated BRAF V600E vs. normal BRAF (titration 50, 10, 1, 0.1, and 0.01 % and NTC in duplicate). Results presented from Fig. 19

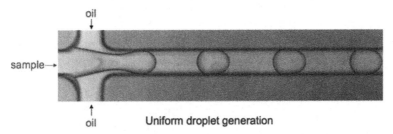

Uniform droplet generation

Fig. 22 Uniform droplet formation. To take advantage of Poisson statistics, partitions (droplets here) must be uniform in size and of known volume. Droplet digital PCR uses intersecting microchannels to create a steady flow of reaction mix that is "pinched" by oil

their reuse will inevitably cause substandard results (if they work at all).

7. Always place samples in the cartridge first, and then add the droplet-generating oil.

8. Uniform-sized droplets are generated by a microfluidic "pinching" effect by the oil when the reaction mix intersects it (Fig. 22).

9. When transferring droplets from the droplet generator to a 96-well plate, use a P50 pipettor (or 8-well multichannel P8 × 50). Using a P50 permits gentle collection and expulsion of the droplets due to the larger piston range that a P50 has vs. a P200.

10. When analyzing a gradient, if the temperature is too high, the partitioned reactions may not amplify and look like wells A06,

B06, and C06 (Fig. 8). As the temperature reaches a more appropriate range, the amplitude of the sub-partition will increase as the PCR reactions become more and more efficient. Most assays have a span of temperatures where results will be the same. As we approach this range, we can sometimes see "stragglers" where some sub-partitions have not fully amplified and are trying to catch up (droplets in wells E06 and F06 of Fig. 8). Care must also be taken not to go too low in temperature annealing where, as with regular PCR, primer specificity is reduced and overall signal (specific product generated) will also drop. This will also create "stragglers," but this time probably due to mispriming and component depletion. This may cause the under-representation of the true amount of target molecules in the sample.

11. If many "stragglers" are present in the temporal (event) or 2D plots, consider increasing the number of cycles by 5–50. Avoid, if possible, excessive cycling beyond 55 cycles.

12. When selecting reaction conditions for a duplex assay, choose a temperature where both assays show good separation of the positive and negative sub-partitioned PCR reactions. Note that both individual assays should work independently of the presence or the absence of the other.

13. Duplex 2D plots typically have an orthogonal distribution (Fig. 23). Migration of the FAM positive subreactions (blue dots in figure) towards the right or HEX positive subreactions (green dots) upwards can indicate cross-reactivity of the PCR assays. This is rare but may indicate that reaction conditions such as primer and probe concentrations are excessive or

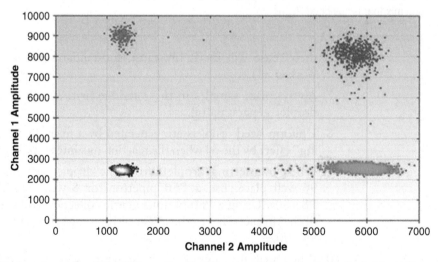

Fig. 23 2D view of a duplex digital PCR reaction. Partitioned sub-assays are typically distributed in an orthogonal pattern

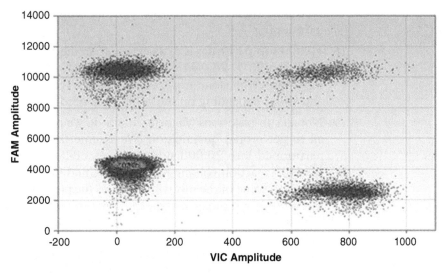

Fig. 24 Non-orthogonal distribution in a duplex assay. Non-orthogonal distribution can be caused by improper fluorophore selection and by cross-reactive assays

that the assays interfere with one another. Non-orthogonal distributions can also be caused by incorrect dye selection in the analysis software. For example in Fig. 24 a HEX-labeled probe was used in conjunction with a FAM probe but software was told that the assay used VIC- and FAM-labeled probes. This can easily be corrected by resetting the analysis parameters.

14. Limit of detection determination requires proper statistical validation and will depend on the background "noise" level of the target one is interested in and the quality of the assay. Noise levels are generally caused by assay amplification criteria (specificity) and by background contamination levels in the lab where the assay is performed. These need to be routinely redetermined.

15. Select a restriction enzyme that will not cut within any of the amplicons of interest.

16. Genomic DNA concentrations greater than 3 ng per μl of PCR reaction create an environment that makes the DNA very viscous and affects proper random distribution across the sample. This will affect the proper quantitation. Concentrated samples also display primer-to-target accessibility issues often generating more "stragglers."

17. When using low-salt buffers for the restriction digest, their inhibitory effect in the PCR reaction will be minimal. We often use volumes of 25 % unclean digested sample in PCR reactions with good positive-to-negative ratios and good quantitative results. Some restriction enzymes can be directly added to the PCR reaction mix (5 U per 20 μl reaction). The reaction is

then incubated for 10–15 min at room temperature prior to partitioning.

18. 10,000 sub-partitions is a "sweet spot" in digital PCR. At this level, the 95 % Poisson confidence interval is a few percent. Pipetting error alone can easily contribute errors of 5–10 % and above depending on the volume pipetted.

19. Not all sub-partitions will be used nor do we need to use them all to get proper quantitation. We assume that 20 μl will be partitioned into 20,000 sub-partitions of 1 nl, but some of the PCR reaction volume will be left behind in the setup tube (or well), some of the mix will be lost in the pipet, some will be lost in the partitioning system, and some will be excluded from analysis due to quality control parameters. Since digital PCR results are determined using a ratio of positive to total events whether we analyze 10,000 events or 20,000 events, the ratios will remain almost the same; thus, the results will remain the same. Random distribution at these partition levels will mean that the events that are unaccounted for will have the same distribution as the ones that are counted. The number of partitions analyzed will have an effect on confidence intervals of the result, with the higher the number of events analyzed, the smaller the 95 % confidence intervals.

20. Results are provided in the form of copies per μl. Although not obvious at first look, there is logic in this reporting process. In reactions that contain large amounts of DNA or in reactions that require higher resolution, the PCR reaction volume may be increased and distributed across multiple wells. For example 60 μl of a PCR reaction are prepared (instead of 20 μl) and split into three dPCR reactions. Due to the digital nature of dPCR and the fact that the three reactions came from the same pool, the results of the three wells can be merged into a single well. This allows the larger samples to be analyzed and better resolution (smaller 95 % confidence intervals) (Fig. 25). When calculating the number of target molecules in this large reaction we would multiply the ddPCR result by 60 $(3 \times 20 \ \mu l)$.

21. Duplex PCR reactions are preferred in dPCR for CNV assays, as inter-sample loading variability is minimized by probing for both targets in the same tube. Multiplex reactions generally work very well in digital PCR due to target compartmentalization that produces a synthetic enrichment effect.

22. One important aspect for proper CNV estimates is sample preparation. Due to the high partitioning level of dPCR reactions and the digital nature of the results (positive or negative for each partition), the sample must be properly digested prior to processing. Unless the potential target gene of interest is known to be spatially distant from the other copies, as in the case with

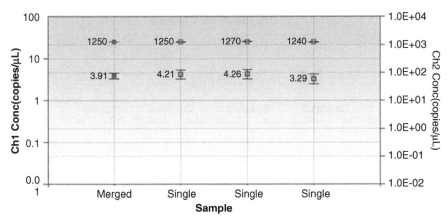

Fig. 25 Example of a merged well analysis. The *leftmost* well is a "merged well" composed of the total positive and negative partitions of the three following individual replicates

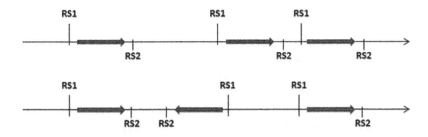

Digestion strategy used	Copy Number Top	Copy Number Bottom
Actual Copy number	3	3
No Digest	1	1
Single restriction enzyme digest RS1	3	2
Single restriction enzyme digest RS2	3	2
Double restriction enzyme digest	3	3

Fig. 26 Hypothetical replicate gene layouts and the effect restriction digest strategy has on determining the correct value. Samples with proximal copies of the target of interest should be separated from one another prior to use in digital PCR. Non-separated linked copies will co-migrate onto single compartments and contribute to an overall Poisson count of one or two vs. the three that are actually present in the sample

trisomies, samples that are either undigested or improperly digested run the risk of having multiple replicate copies migrate into individual partitions and counted as single copies (counted as a single positive event). Proper digestion allows for random distribution of the target and reference molecules in a pattern that is amenable to Poisson statistics (Fig. 26).

23. Comparing a sample's digested and undigested copy number can help determine whether the replicate copies of the gene are

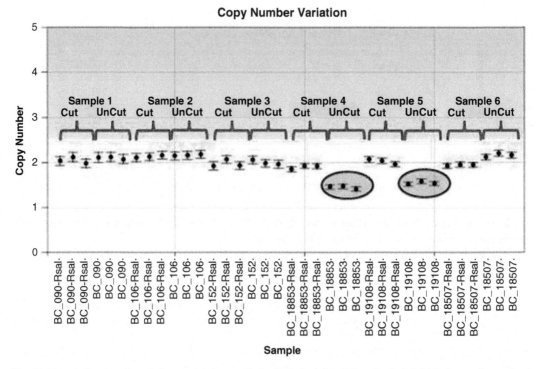

Fig. 27 Digested and undigested copy number analysis of six Coriell cell lines in digital PCR. Comparison of cut and uncut samples can be used to determine proximity of replicate genes

distal or proximal if one chooses to perform this analysis. If a sample has two copies of a target gene located far from one another or on separate chromosomes, digested and undigested copy number analysis will yield the same value. If the two molecules are proximal the undigested sample will have a lower copy number that is expected due to the co-migration of the two copies into the same sub-partition and being counted as one instead of being randomly distributed in possibly two sub-partitions of the dPCR reaction and being counted as two (once again following Poisson distribution statistics). The results will rarely if ever be exactly half, as random shearing of the DNA molecules during sample extraction and subsequent handling will inevitably contribute to the separation of some proximal replicate copies (Fig. 27).

24. As with quantitative assays, increasing the number of partitions increases the resolution and reduces confidence intervals. For assays requiring high levels of discrimination the sample can be run in multiple wells and the results combined in a merged well (Fig. 28).

25. In theory one would like to input as much DNA as possible into a single 20 μl reaction to have the highest detection capability possible. In real life the amount of DNA used will depend

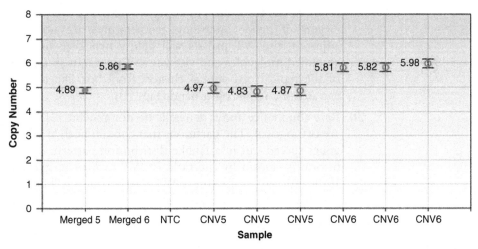

Fig. 28 Merged CNV data. For increased resolution, technical replicate samples can be combined. *Left* two wells are the merged equivalent of three replicates each (trailing wells). Error bars represent 95 % confidence intervals

on factors such as the ploidy of the sample, size of the genome, and concentration of the target within the sample.

26. The dynamic range of the assay depends upon the partition number; typically, five to six times the number of partitions generated (100,000–120,000 target molecules for a 20,000 partition assay) exceeding this number of input molecules will not permit for accurate quantitation of WT molecules. If one is studying a tetraploid or hexaploid sample, this must be considered during experimental setup. Running parallel assays by generating a larger reaction (40, 60, 80 μl) and merging the data (into a metawell) is also possible to increase the dynamic range of the assay.

27. Highly concentrated DNA samples should be sheared or digested to ensure easy sample distribution and maintain fluidity of the PCR reaction. Typically concentration levels above 60 ng per 20 μl should be digested.

28. Depending on the source material used and the abundance of the mutation one is looking for, there are physical limits as to how much sample can be extracted from an organism (and maintain viability). Although detection of 1 in 100,000 or 1 in a million sounds good in theory, is it practical? One must remember that at low detection levels errors attributable to subsampling (e.g., chance of collecting 1 or 2 target molecules in 5 ml of blood) and sample preparation become a greater variable to detection and quantitation than do current instrumentation limitations.

29. The strange look of mutation abundance temporal (event) plots is often due to cross-reactivity of probes within the assay.

As there can be as little as a single base between WT and Mut targets, if probe annealing temperatures are not carefully adjusted, the WT probes will lightly hybridize on the Mut amplified molecules and vise versa. This results in a migration of the typical orthogonal clusters from box like to swept in towards the upper right hand (Fig. 19).

30. Rare abundance assays generally do not generate 2D plots that are orthogonal. The partitions that contain both WT and Mut genes spread out in a fanlike distribution pattern. This distribution is caused by the ratio of WT to Mut target present. If the ratio is 1 to 1 (Mut to WT) the partition will be closer to the Mut cluster, while a more dilute partition (say 1 to 4) will present itself closer to the WT partition (Fig. 19).

Acknowledgements

I would like to acknowledge and thank my colleagues at Bio-Rad Laboratories, Adam McCoy, George Karlin-Neumann, Jack Reagan, and Svilen Tzonev, and all the members of the Digital Biology Center for their contributions to this overview and for their tireless efforts in the development and advancement of digital PCR.

References

1. Huggett JF, Foy CA, Benes V et al (2013) The digital MIQE guidelines: minimum information for publication of quantitative digital PCR experiments. Clin Chem 59:892–902

2. Strain MC, Richman DD (2013) New assays for monitoring residual HIV burden in effectively treated individuals. Curr Opin HIV AIDS 8:106–110

3. Hatano H, Strain MC, Scherzer R et al (2013) Increase in 2-LTR circles and decrease in D-dimer after raltegravir intensification in treated HIV-infected patients: a randomized, placebo-controlled trial. J Infect Dis 208:1436–1442

4. Strain MC, Lada SM, Luong T et al (2013) Highly precise measurement of HIV DNA by droplet digital PCR. PLoS One 8:e55943. doi:10.1371/journal.pone.0055943

5. Kelly K, Cosman A, Belgrader P et al (2013) Detection of methicillin-resistant Staphylococcus aureus by a duplex droplet digital PCR assay. J Clin Microbiol 51:2033–2039

6. Boettger LM, Handsaker RE, Zody MC et al (2012) Structural haplotypes and recent evolution of the human 17q21.31 region. Nat Genet 44:881–885

7. Gevensleben H, Garcia-Murillas I, Graeser MK et al (2013) Noninvasive detection of HER2 amplification with plasma DNA digital PCR. Clin Cancer Res 19:3276–3284

8. Hindson BJ, Ness KD, Masquelier DA et al (2011) High-throughput droplet digital PCR system for absolute quantitation of DNA copy number. Anal Chem 83:8604–8610

9. Nadauld L, Regan JF, Miotke L et al (2012) Quantitative and sensitive detection of cancer genome amplifications from formalin fixed paraffin embedded tumors with droplet digital PCR. Transl Med (Sunnyvale). 2. doi:pii: 1000107

10. Yeh I, von Deimling A, Bastian BC (2013) Clonal BRAF mutations in melanocytic nevi and initiating role of BRAF in melanocytic neoplasia. J Natl Cancer Inst 105:917–919

Chapter 17

Developing Noninvasive Diagnosis for Single-Gene Disorders: The Role of Digital PCR

Angela N. Barrett and Lyn S. Chitty

Abstract

Cell-free fetal DNA constitutes approximately 10 % of the cell-free DNA found in maternal plasma and can be used as a reliable source of fetal genetic material for noninvasive prenatal diagnosis (NIPD) from early pregnancy. The relatively high levels of maternal background can make detection of paternally inherited point mutations challenging. Diagnosis of inheritance of autosomal recessive disorders using qPCR is even more challenging due to the high background of mutant maternal allele. Digital PCR is a very sensitive modified method of quantitative real-time PCR (qPCR), allowing absolute quantitation and rare allele detection without the need for standards or normalization. Samples are diluted and then partitioned into a large number of small qPCR reactions, some of which contain the target molecule and some which do not; the proportion of positive reactions can be used to calculate the concentration of targets in the initial sample. Here we discuss the use of digital PCR as an accurate approach to NIPD for single-gene disorders.

Key words Digital PCR (dPCR), Microfluidics, Quantitative real-time PCR, Hydrolysis probes, Noninvasive prenatal diagnosis, Cell-free fetal DNA

1 Introduction

Digital PCR, first described by Vogelstein in 1999 [1], is a highly sensitive technique which involves the partitioning of a single sample into many individual PCR reactions at limiting dilution, allowing the total number of copies of a target molecule in the initial undiluted sample to be determined. This can facilitate the detection of rare variants in a high background of wild-type sequences, such as those found in cancers [2–4], and cell-free fetal DNA (cffDNA) [5–7], or in microbiological applications, for example viral detection [8, 9] and detection of pathogens such as methicillin-resistant *Staphylococcus aureus* (MRSA) [10]. Point mutations with frequencies as low as 1 in 100,000 have been detected using droplet digital PCR (ddPCR) [11]. Digital PCR is far superior to quantitative real-time PCR (qPCR) for detecting small percentage of changes in copy numbers; at best qPCR can measure 1.25–1.5-fold changes in copy number [12], whereas dPCR can measure less than 1.2-fold

Roberto Biassoni and Alessandro Raso (eds.), *Quantitative Real-Time PCR: Methods and Protocols*, Methods in Molecular Biology, vol. 1160, DOI 10.1007/978-1-4939-0733-5_17, © Springer Science+Business Media New York 2014

[13] and has been successfully used for detection of change in *HER2* copy numbers in breast cancer samples [14–16].

Digital PCR is performed either as a real-time reaction or as an endpoint reaction depending on the technology platform used, but for all methods the sample must be partitioned in some way; the dilution of the sample is such that some of these partitions will contain no target molecule, giving a negative result, whilst others will contain one or more target molecules, thus giving a positive result. In samples with a high dilution, the number of positive wells is equal to the number of target molecules. As the sample becomes more concentrated, in order to account for the fact that more than one molecule may be present in a positive partition, a Poisson correction is applied to estimate actual target molecules present in the reaction, and then this combined with the volume of the sample assayed is used to calculate the absolute concentration of target DNA molecules [17]. Since digital PCR relies on a binary output—present or absent—even assays with relatively poor amplification efficiency can be used to determine copy numbers [18].

The first digital PCR experiments were carried out using 96-well plates [1], but a large number of plates were required to gain sufficient data for analysis, and so this approach was costly and impractical. In 2006, Fluidigm introduced the BioMark™, the first commercial system for digital PCR, which is based on microfluidics. Nanofluidic chips can be used to analyze several samples in parallel; for example, the 12-panel Fluidigm Digital Array integrated fluidic circuit (IFC) allows up to 12 samples to be analyzed at a single time, giving 765 partitions per sample. If a higher throughput is needed, ddPCR methods may be of more use, making use of emulsion PCR to form up to 20,000 partitions per sample, but these are yet to be extensively evaluated in clinical practice [11].

One application for which digital PCR has been shown to have particular promise is in the area of noninvasive prenatal diagnosis (NIPD). Prenatal diagnosis is an established part of obstetric practice, and genetic diagnosis is offered to women at high risk of carrying a fetus with a single-gene disorder or aneuploidy. Currently most definitive prenatal diagnosis requires fetal material which is obtained using an invasive test (chorionic villus sampling (CVS) or amniocentesis). These procedures carry a small but significant risk of miscarriage of around 0.5–1 % [19] and cannot be performed until after 11 weeks of gestation [20].

A major research goal in prenatal diagnosis has been to develop methods to enable diagnosis to be carried out without the risk of miscarriage by using fetal genetic material circulating in the maternal blood. Early work focused on the isolation and analysis of fetal cells in the maternal circulation, but more recently this has moved to analysis of cell-free fetal DNA (cffDNA) following its identification in maternal plasma in the late 1990s [21]. cffDNA, which originates from the placenta [22] and circulates alongside cell-free maternal DNA in maternal plasma from as early as 5 weeks gestation [23], constitutes around 10 % of the total cell-free DNA (cfDNA) [24], and is made up of small fragments with an average length of 143 bp, 20 bp less than maternal DNA, which has an average length of 166 bp [25]. It is rapidly cleared from maternal circulation with a very short half-life of 16 min, so that it is usually undetectable just 2 h after birth [26].

Noninvasive prenatal testing (NIPT) is already a reality for some indications such as fetal sex determination in women at risk of X-linked disorders or congenital adrenal hyperplasia [27] and the testing of Rhesus D (RhD)-negative women at risk of hemo-lytic disease of the fetus and newborn (HDFN) [28]. The use of NIPT for Down's syndrome diagnosis was first reported in 2008 [29, 30]. Over the last 2 years several large-scale validation studies have since demonstrated high sensitivities and specificities using both massively parallel and targeted sequencing approaches [31]. NIPT for Down's syndrome is now available commercially in several countries [32].

There has been less emphasis on developing NIPD for single-gene disorders, largely because these tend to require development on a patient- or a disease-specific basis. Initially studies focused on identifying mutations that have been inherited paternally or arise de novo using a wide variety of PCR-based techniques (reviewed by Lench et al. [33]). Proof-of-principle studies using digital PCR have been published demonstrating the potential for the diagnosis of a number of autosomal recessive single-gene disorders where both parents carry the same mutation, including β-thalassaemias [5] and sickle cell anaemia [6], or where the mutation is inherited maternally, as in the case of haemophilia [7]. Diagnosis in these situations requires accurate estimation of allelic ratios, which can be done using a quantitative approach known as relative mutation dosage (RMD) [5]. If a woman is heterozygous and the fetus is also a heterozygote for the same mutant allele it is expected that there will be an allelic balance between the wild-type and mutant alleles; if the fetus is homozygous for either the mutant or the wild-type allele there will be an overrepresentation of one or the other, which can be assessed statistically using sequential probability ratio testing (SPRT). RMD is dependent upon accurate assess-ment of fractional fetal DNA concentration, which can readily be determined in male fetuses using sequences on the Y chromo-some, but in cases where the fetus is female it is reliant on the detection of paternally inherited SNPs or insertion/deletion (indel) polymorphisms [6].

More recently it has been shown that digital PCR can be used as a more sensitive approach to qPCR for identification of paternally inherited mutations [33].

Whilst digital PCR is far superior to qPCR in terms of sensitivity, there are limitations and disadvantages to this technique as well. Designing probes for a family-specific mutation is relatively costly, and, because of the need to run several controls simultaneously, it is really only possible to run a maximum of two assays on a single digital array, further increasing costs. Thus, when looking for known mutations, provided the assay is developed, application of dPCR can be relatively straightforward. However, when screening a gene for multiple possible mutations serial assays would be required, increasing costs and time to deliver a result. Droplet digital PCR allows for a far greater number of samples to be run at one time, and there are a greater number of partitions, allowing for quantitation of between 1 and 100,000 target molecules, but this requires an increased number of technical steps, bringing a new set of technical challenges [9].

Here we describe in detail the process of developing a digital PCR assay for the detection of a paternal or a de novo mutation from a maternal plasma cfDNA sample.

2 Materials

2.1 Equipment and Consumables

1. Microtubes (1.5 ml).
2. 0.2 ml thin-walled PCR strip tubes and caps.
3. MicroAmp optical 96-well plates (Life Technologies).
4. Optical plate seals (Life Technologies).
5. Real-time PCR system.
6. Vortex mixer.
7. Microcentrifuge.
8. Plate centrifuge.
9. Fluidigm BioMark™.
10. MX IFC Controller (Fluidigm).
11. 10, 20, 200, and 1,000 μl filter tips.
12. 10, 20, 200, and 1,000 μl pipettes.

2.2 Quantitative Real-Time PCR Reagents

1. Taqman Universal PCR Mastermix (no amperase UNG; Life Technologies).
2. Forward and reverse primers (Sigma Aldrich).
3. FAM-MGB and VIC-MGB labeled probes (Life Technologies).
4. Nuclease free water (Sigma Aldrich).

2.3 Digital PCR Reagents

1. 12.675 Digital Array IFC (Fluidigm).
2. Control Line Fluid (Fluidigm).
3. Gene Expression Master Mix (Life Technologies).
4. GE sample loading buffer (Fluidigm).
5. FAM-MGB and VIC-MGB labeled probes (Life Technologies).
6. Forward and reverse primers (Sigma Aldrich).

3   Methods

3.1   Design of Primers and Probes

Digital PCR primers are designed using Primer 3 software [34] following the same standard guidelines as for those used for design of qPCR primers (*see* **Note 1**). For NIPD we are assaying short fetal DNA fragments, and therefore it is important to design the amplicons to be as short as possible [35], with the ideal length being under 100 bp. Following design of primers, two allele-specific Taqman-MGB hydrolysis probes (*see* **Note 2**), one for the wild-type allele (labeled with FAM) and the other for the mutant allele (labeled with VIC), should be designed using dedicated software (i.e., Primer Express (Applied Biosystems) or similar; *see* **Note 3**).

3.2   Preparation of 20× Duplex Primer/Probe Mixes

1. Dilute primers to 100 µM using nuclease-free H_2O, and further dilute probes to 20 µM by adding 20 µl of stock probe to 80 µl of H_2O in a 1.5 ml tube, vortexing well.

2. Make a 20× duplex primer/probe stock mix as seen in Table 1.

3. Vortex well, and centrifuge to spin down. Store at –20 °C in aliquots or at 4 °C for up to 4 weeks.

3.3   Quantitative Real-Time PCR to Confirm Assay Specificity

Ideally genomic DNA samples are required to confirm specificity of an allelic discrimination assay (*see* **Note 4**). One sample should be homozygous for the wild-type allele, and the other sample should be heterozygous.

1. Dilute gDNAs to 5 ng/µl each in nuclease-free H_2O.

2. Make a qPCR mastermix in a 1.5 ml tube as seen in Table 2.

Table 1
Preparation of 20× duplex primer/probe mixes for digital PCR

		Concentration in 20× mix (µM)
Forward primer (100 µM)	45 µl	18
Reverse primer (100 µM)	45 µl	18
FAM probe (20 µM)	50 µl	4
VIC probe (20 µM)	50 µl	4
H_2O	60 µl	–

Table 2
Quantitative real-time PCR mix to confirm specificity of the duplex assays

Taqman universal PCR mastermix (no amperase UNG)	90 µl
20× Duplex primer/probe assay	9 µl
H_2O	36 µl

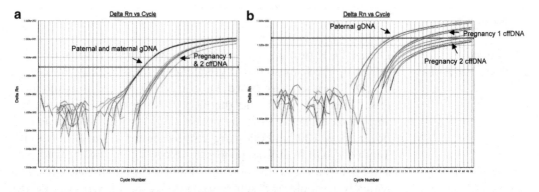

Fig. 1 Quantitative real-time PCR curves produced during validation of an assay for Fraser syndrome. (**a**) Both the maternal and paternal gDNA samples, as well as the cfDNAs from the first and second pregnancies, are homozygous for the wild-type allele (c.10216C). (**b**) The paternal gDNA is heterozygous for the c.10216C>T mutation, as is a cfDNA sample from their first affected pregnancy, and the maternal gDNA and second pregnancy cfDNA sample are negative

3. Pipette 15 μl of mastermix into nine wells of a 96-well optical plate.

4. Add 5 μl of wild-type gDNA to the first three wells, 5 μl of heterozygous gDNA (or cfDNA) to wells four to six, and H$_2$O as a no-template control (NTC) in wells seven to nine.

5. Seal the plate using an optical plate seal. Smooth down edges carefully to prevent evaporation during thermal cycling.

6. Vortex gently and centrifuge plate to spin down droplets. Ensure that there are no bubbles present at the bottom of the wells. If there are, "snap" the wells with the bubbles from the bottom to remove them and re-centrifuge if necessary.

7. Transfer the plate to the real-time qPCR machine, and perform qPCR with the following thermocycling conditions: 95 °C for 10 min followed by 45 cycles of 95 °C for 15 s and 60 °C for 60 s.

8. Examine amplification curves to confirm the assay specificity. It is expected that the wild-type gDNA should give positive replicates only for the wild-type assay; the heterozygous gDNA or cfDNA should have a positive amplification curve for both the wild-type and mutant assays; and the NTC should be negative for both assays. If this is the case, you can proceed to digital PCR (Fig. 1).

3.4 Digital PCR

Prime Digital Array IFC

1. Switch on the MX IFC Controller.

2. Take a Digital Array IFC, and carefully load control line fluid into the array into the control line fluid inlets on both sides (*see* **Note 5**).

3. Place the array into the MX IFC Controller with barcode facing forwards, and close the tray. Press "Prime chip" followed by "Run script."

Table 3
Components of the digital PCR mastermix

Gene expression mastermix	65 µl
Duplex assay	6.5 µl
GE sample loading buffer	6.5 µl

Mutation Detection in
cfDNA Using Digital PCR

There are 12 inlets in a 12.765 digital array, in which you can run up to 12 samples. Since it is ideal to run samples in duplicate, the arrangement can be as follows: duplicate panels for both parent's gDNA, a cfDNA or a gDNA sample from a previously affected pregnancy, the cfDNA sample from the current pregnancy, a cfDNA sample from an unrelated unaffected pregnancy, and NTCs.

1. Mix the digital PCR mastermix in a 1.5 ml tube (*see* Table 3).

2. Label six 0.2 ml PCR tubes, and pipette 12 µl of mastermix into each.

3. Add 8 µl of each cfDNA sample to each tube.

4. Vortex tubes thoroughly and centrifuge to spin down droplets.

5. Remove the primed IFC from the MX IFC Controller, and close the tray.

6. Load 10 µl of H_2O into the two outer hydration wells of the chip (marked H).

7. Carefully load 9.5 µl of the PCR reaction mix into each of the first two wells (numbered one and two), ensuring that there are no bubbles (*see* **Note 6**).

8. Repeat this in the remaining wells with duplicates of each sample.

9. Return the chip to the MX IFC Controller, and press "load," followed by "run script."

10. When loading has finished, remove the Digital IFC Array from the MX IFC Controller.

11. Remove the blue sticker from the bottom of the chip, and use tape to remove dust from the top surface of the Digital IFC Array if necessary (*see* **Note 7**).

12. Open BioMark Collection Software, and click "start new run." Load the IFC onto the tray with the barcode of the IFC facing outwards.

13. Click "Load" and then "Next."

14. Name the IFC, and select "Next."

15. Select two probes, FAM-MGB for probe 1 and VIC-MGB for probe 2.

16. Use "default 45 cycles" as the thermal protocol, which uses the following conditions: 50 °C for 2 min and 95 °C for 10 min, followed by 45 cycles of 95 °C for 15 s and 60 °C for 1 min.

17. Click "Start Run" (*see* **Notes 8** and **9**).

Data Analysis

1. When the run has finished open the Fluidigm Digital PCR Analysis software.

2. Under "Quick tasks" select "Open a chip run" and then navigate to the ChipRun.bml file that you want to open (*see* **Note 10**).

3. Go to "panel summary." In analysis settings keep "quality threshold" at 0.65. Change target Ct for FAM and VIC to 45. The Ct threshold method should be changed to "user global." Click "Analyze."

4. Adjust the thresholds as necessary (*see* **Note 11**). Thresholds will vary and should be set at a point where the qPCR curves are in the exponential phase.

5. If any curves look abnormal, click on "Panel details," go to "Heatmap View," and highlight individual channels in the panel to identify the wells producing abnormal curves. Fail any that look abnormal.

6. When all curves for the FAM and the VIC signals appear acceptable, click File, Export, and save as type "Summary table results.csv" and examine the predicted target molecules for each sample.

7. For an unaffected sample we would expect to see only wild-type targets; for a heterozygous gDNA there should be roughly equal counts for wild type and mutation. In the case of an affected fetus, the cfDNA should produce a high number of wild-type targets (originating from the maternal cfDNA) and a lower percentage of mutant counts (from the fetus, inherited paternally or occurring de novo).

An example is shown for a family with the autosomal recessive disorder, Fraser syndrome, caused by mutations to the *FRAS1* gene (Fig. 2). The mother has a large deletion encompassing exons 50–71, whilst the father has a point mutation at position c.10261C>T. The first pregnancy from this couple was affected with Fraser syndrome, and so the second pregnancy was also tested using the noninvasive test designed to exclude the paternal mutation. As shown in Fig. 2, the father is positive for both wild-type and mutant alleles, as is the cfDNA sample from the first affected pregnancy. The second pregnancy and an unrelated cfDNA sample from a different woman are both unaffected.

a

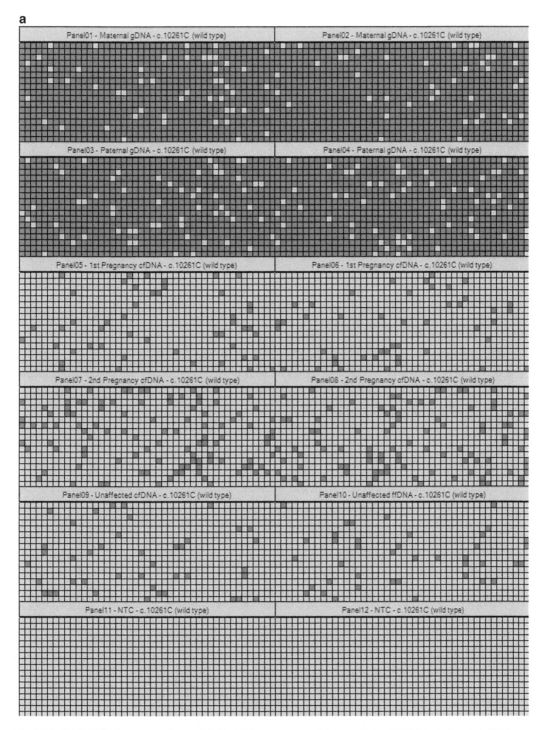

Fig. 2 Digital PCR for Fraser syndrome. (**a**) *Red dots* represent wild-type alleles (FAM-labeled probe). Wild-type signals are present in all samples. (**b**) *Blue dots* represent mutant alleles (c.2016C>T; VIC-labeled probe), which are only seen in the paternal gDNA (panels *3* and *4*), and cfDNA sample from the first affected pregnancy (panels *5* and *6*). The second pregnancy was unaffected (panels *7* and *8*), as was a normal control cfDNA from an unrelated family (panels *9* and *10*) (adapted from Lench et al. [33])

b

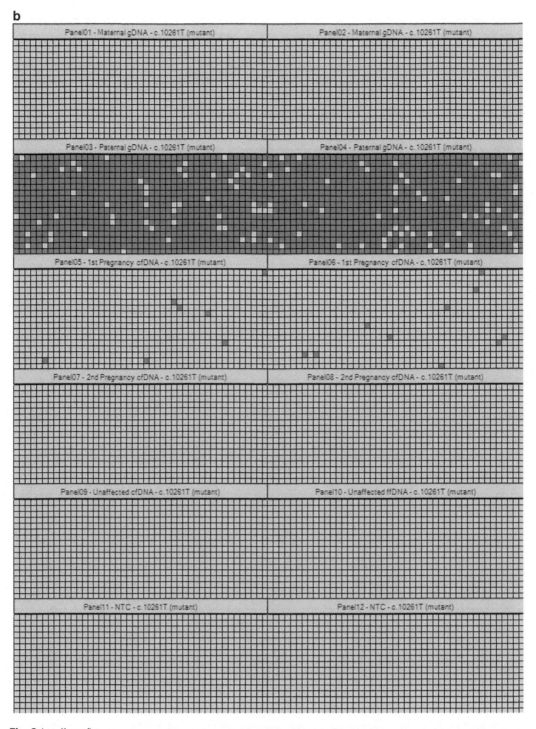

Fig. 2 (continued)

4   Notes

1. Standard primer design requires adherence to the following guidelines:

   (a) Avoid runs of four or more identical nucleotides.

   (b) The estimated melting temperature (T_m) should be 58–60 °C.

   (c) GC content should be no more than 65 %.

   (d) The last five nucleotides at the 3′ end of each primer should contain no more than two G or C residues.

2. Minor groove binding (MGB)-labeled probes are ideal for use for allelic discrimination for two reasons: firstly, the fact that they use a non-fluorescent quencher allows the reporter dye contribution to the signal to be measured more accurately, and secondly the MGB moiety increases the melting temperature of the probes, allowing design of shorter, more specific probes that can discriminate well between a single mismatched base.

3. Primer Express probe design guidelines are as follows:

   (a) Position the polymorphic site towards the middle of the probe.

   (b) Avoid placing a guanine residue at the 5′ end of the probe, since a guanine adjacent to the reporter dye will quench the reporter fluorescence, weakening the signal.

   (c) The T_m should be 65–67 °C.

   (d) The probe length should be as short as possible (but more than 13 bp in length).

4. If looking for a paternally inherited dominant disorder, a hetero-zygous gDNA sample can be obtained from the father. If the disorder is de novo, there may not be a gDNA sample, and therefore a cfDNA sample from a previously affected preg-nancy is a possible source of heterozygous DNA for validation of the assay.

5. Invert the syringe when removing the cap to prevent the oil solution dripping out before the plunger is depressed. Hold in an inverted position, position the array at an approximately 45° angle, and then position the syringe in the appropriate holes. If any oil does spill over onto the array, use lint-free tissue to remove it immediately.

6. If there are tiny bubbles present at the edges of the well, or they are floating in the liquid, this does not present a problem, but if the bubbles are large or positioned over the center of the well, the whole solution needs to be carefully aspirated with a pipette back into the original tube, centrifuged, and then dispensed with a new pipette tip. Bubbles in the two hydration

(H) wells invariably cause dehydration of the chip and must also be avoided.

7. Do not wipe the surface of the chip with tissues to remove dust, as this will only increase the problem.

8. Usually small pieces of debris can be removed with tape. Use a piece of tape of about 5–6 cm. Place the tape firmly over the top surface of the IFC, use a finger to smooth out the tape, and then remove it in one swift action. The tape is about 1/3 the width of the IFC surface, so you will need to repeat a total of three times.

9. It is essential that once the Fluidigm BioMark run is started, the computer is not used for any other purpose as this can lead to crashing of the software.

10. When opening the file to look at the analyzed data, one will need the ChipRun.bml file and the ChipRun.processed.bin (for real-time data). To analyze the data from scratch on another PC, one will need the original ChipRun.bml file (or if previously analyzed, the ChipRun.bml.orig) and the Data folder.

11. With panels set to "User Global," all samples will have the same VIC threshold and all samples will have the same FAM threshold. Start by examining curves at a threshold of 0.1 for both FAM and VIC. If no curves are visible, lower the threshold to 0.05 and then click "analyze." This should be repeated until the curves are all visible, except in the NTC samples, which should be negative. If it is not possible to use a single threshold for all panels with the same fluorescent label, then change the Target Threshold Method to "User Panel" and set each one individually.

The process for analyzing autosomal recessive disorders where both parents carry the same mutation, or for X-linked disorders, would be similar with regard to design of the primers and probes and for the priming and loading of the Digital Array IFCs, but would require an additional Digital Array IFC to be run to quantify the fractional fetal DNA concentration [5–7] and the allelic ratios would be analyzed using SPRT analysis as previously described.

References

1. Vogelstein B, Kinzler KW (1999) Digital PCR. Proc Natl Acad Sci U S A 96:9236–9241

2. Azuara D, Ginesta MM, Gausachs M et al (2012) Nanofluidic digital PCR for KRAS mutation detection and quantification in gastrointestinal cancer. Clin Chem 58:1332–1341

3. Wang J, Ramakrishnan R, Tang Z et al (2010) Quantifying EGFR alterations in the lung cancer genome with nanofluidic digital PCR arrays. Clin Chem 56:623–632

4. Yung TK, Chan HC, Mok TS et al (2009) Single-molecule detection of epidermal growth factor receptor mutations in plasma by microfluidics digital PCR in non-small cell lung cancer patients. Clin Cancer Res 15: 2076–2084

5. Lun FM, Tsui NB, Chan KC et al (2008) Noninvasive prenatal diagnosis of monogenic diseases by digital size selection and relative mutation dosage on DNA in maternal plasma. Proc Natl Acad Sci U S A 105:19920–19925

6. Barrett AN, McDonnell TC, Chan KC et al (2012) Digital PCR analysis of maternal plasma for noninvasive detection of sickle cell anemia. Clin Chem 58:1026–1032

7. Tsui NB, Kadir RA, Chan KC et al (2011) Noninvasive prenatal diagnosis of hemophilia by microfluidics digital PCR analysis of maternal plasma DNA. Blood 117:3684–3691

8. Henrich TJ, Gallien S, Li JZ et al (2012) Low-level detection and quantitation of cellular HIV-1 DNA and 2-LTR circles using droplet digital PCR. J Virol Methods 186:68–72

9. Strain MC, Lada SM, Luong T et al (2013) Highly precise measurement of HIV DNA by droplet digital PCR. PLoS One 8:e55943. doi:10.1371/journal.pone.0055943

10. Hua Z, Rouse JL, Eckhardt AE et al (2010) Multiplexed real-time polymerase chain reaction on a digital microfluidic platform. Anal Chem 82:2310–2316

11. Hindson BJ, Ness KD, Masquelier DA et al (2011) High-throughput droplet digital PCR system for absolute quantitation of DNA copy number. Anal Chem 83:8604–8610

12. Weaver S, Dube S, Mir A et al (2010) Taking qPCR to a higher level: analysis of CNV reveals the power of high throughput qPCR to enhance quantitative resolution. Methods 50:271–276

13. Whale AS, Huggett JF, Cowen S et al (2012) Comparison of microfluidic digital PCR and conventional quantitative PCR for measuring copy number variation. Nucleic Acids Res 40:e82. doi:10.1093/nar/gks203

14. Belgrader P, Tanner SC, Regan JF et al (2013) Droplet digital PCR measurement of HER2 copy number alteration in formalin-fixed paraffin-embedded breast carcinoma tissue. Clin Chem 59:991–994

15. Gevensleben H, Garcia-Murillas I, Graeser MK et al (2013) Noninvasive detection of HER2 amplification with plasma DNA digital PCR. Clin Cancer Res 19:3276–3284

16. Heredia NJ, Belgrader P, Wang S et al (2013) Droplet digital PCR quantitation of HER2 expression in FFPE breast cancer samples. Methods 59:S20–S23. doi:10.1016/j.ymeth.2012.09.012

17. Dube S, Qin J, Ramakrishnan R (2008) Mathematical analysis of copy number variation in a DNA sample using digital PCR on a nanofluidic device. PLoS One 3:e2876. doi:10.1371/journal.pone.0002876

18. Zimmermann BG, Dudarewicz L (2008) Real-time quantitative PCR for the detection of fetal aneuploidies. Methods Mol Biol 444:95–109

19. Tabor A, Alfirevic Z (2010) Update on procedure-related risks for prenatal diagnosis techniques. Fetal Diagn Ther 27:1–7

20. Alfirevic Z, Sundberg K, Brigham S (2003) Amniocentesis and chorionic villus sampling for prenatal diagnosis. Cochrane Database Syst Rev 3, CD003252

21. Lo YM, Corbetta N, Chamberlain PF et al (1997) Presence of fetal DNA in maternal plasma and serum. Lancet 350:485–487

22. Alberry MS, Maddocks DG, Hadi MA et al (2009) Quantification of cell free fetal DNA in maternal plasma in normal pregnancies and in pregnancies with placental dysfunction. Am J Obstet Gynecol 200(98):e1–e6. doi:10.1016/j.ajog.2008.07.063

23. Birch L, English CA, O'Donogue K et al (2005) Accurate and robust quantification of circulating fetal and total DNA in maternal plasma from 5 to 41 weeks of gestation. Clin Chem 51:312–320

24. Lun FM, Chiu RW, Allen Chan KC et al (2008) Microfluidics digital PCR reveals a higher than expected fraction of fetal DNA in maternal plasma. Clin Chem 54:1664–1672

25. Lo YM, Chan KC, Sun H et al (2010) Maternal plasma DNA sequencing reveals the genome-wide genetic and mutational profile of the fetus. Sci Transl Med 2:61ra91. doi:10.1126/scitranslmed.3001720

26. Lo YM, Zhang J, Leung TN et al (1999) Rapid clearance of fetal DNA from maternal plasma. Am J Hum Genet 64:218–224

27. Hill M, Finning K, Martin P et al (2011) Noninvasive prenatal determination of fetal sex: translating research into clinical practice. Clin Genet 80:68–75

28. Daniels G, Finning K, Martin P et al (2009) Noninvasive prenatal diagnosis of fetal blood group phenotypes: current practice and future prospects. Prenat Diagn 29:101–107

29. Chiu RW, Chan KC, Gao Y et al (2008) Noninvasive prenatal diagnosis of fetal chromosomal aneuploidy by massively parallel genomic sequencing of DNA in maternal plasma. Proc Natl Acad Sci U S A 105:20458–20463

30. Fan HC, Blumenfeld YJ, Chitkara U et al (2008) Noninvasive diagnosis of fetal aneuploidy by shotgun sequencing DNA from maternal blood. Proc Natl Acad Sci U S A 105:16266–16271

31. Boon EM, Faas BH (2013) Benefits and limitations of whole genome versus targeted

approaches for noninvasive prenatal testing for fetal aneuploidies. Prenat Diagn 33:563–568

32. Chitty LS, Hill M, White H et al (2012) Non-invasive prenatal testing for aneuploidy-ready for prime time? Am J Obstet Gynecol 206:269–275

33. Lench N, Barrett A, Fielding S et al (2013) The clinical implementation of non-invasive prenatal diagnosis for single-gene disorders:

challenges and progress made. Prenat Diagn 33:555–562

34. Rozen S, Skaletsky HJ (1998) Primer3. Code available at http://www-genome.wi.mit.edu/genome_software/other/primer3.html

35. Sikora A, Zimmermann BG, Rusterholz C et al (2010) Detection of increased amounts of cell-free fetal DNA with short PCR amplicons. Clin Chem 56:136–138

INDEX

Roberto Biassoni and Alessandro Raso (eds.), *Quantitative Real-Time PCR: Methods and Protocols*, Methods in Molecular Biology,
vol. 1160, DOI 10.1007/978-1-4939-0733-5, © Springer Science+Business Media New York 2014

CPSIA information can be obtained
at www.ICGtesting.com
Printed in the USA
LVOW02*1625150117

521019LV00003B/105/P